Tourism Before, During and After Corona

Christian J. Jäggi

Tourism Before, During and After Corona

Economic and Social Perspectives

 Springer

Christian J. Jäggi
Meggen, Switzerland

ISBN 978-3-658-39181-2   ISBN 978-3-658-39182-9   (eBook)
https://doi.org/10.1007/978-3-658-39182-9

Responsible Editor: Claudia Rosenbaum
This Springer imprint is published by the registered company Springer Fachmedien Wiesbaden GmbH, part of Springer Nature.
The registered company address is: Abraham-Lincoln-Str. 46, 65189 Wiesbaden, Germany

Preface

One of the sectors most affected by the Covid-19 pandemic was tourism. This drop was doubly serious after years and decades of constant tourist expansion. But even the most drastic declines in tourism will never turn off people's travel lust for a long time. On the contrary: people are today more mobile than ever before in their history—and this mobility is not only geographical, but also social and above all virtual. At the same time, more and more people will want to travel around the world—growing incomes of the middle class and still increasing population numbers in many countries, especially in Africa and parts of Asia, are important drivers of tourism demand. Conversely, the number of travel destinations is not unlimited—and already today many tourist destinations suffer from overtourism, large numbers of tourists and high pressure on local communities as well as strong environmental impacts. New forms of virtual travel will emerge, but also tourist space flights and finally trips to other planets. On Earth, it will be hardly possible to avoid protecting tourist resources, limiting the extent of tourism locally, temporally and in terms of frequency of travel—too great is the ecological footprint of travel.

Meggen   Christian J. Jäggi
in July 2021

Contents

1	**Introduction**	1
	References	2
2	**Tourism—A Short History**	3
	2.1 To the Term	3
	2.2 Some Numbers and Facts	5
	2.3 Development	10
	2.4 Reasons to Travel	14
	2.5 Travel Destinations	15
	2.6 Tourist Gaze	17
	2.7 Tourism During and After Corona	18
	References	23
3	**Economic Importance of Tourism**	27
	3.1 Global	30
	3.2 National	32
	3.3 Local	34
	3.4 Positive and Negative Impacts of Tourism	35
	3.5 Economic Problems in Tourism	37
	3.6 Tourism generates Jobs	37
	3.7 New Ways of Working	38
	3.8 Tourism as a Cause of Dual Markets and Inflation	39
	3.9 Economic Perspectives for Tourism After Corona	39
	3.10 Tourism and the Economic Effects of Crises—An Endless Story?	42
	References	44
4	**Ecological Consequences of Tourism**	47
	4.1 Mass and Cheap Tourism	50
	4.2 Tourism and Climate Change	51
	4.3 More Environmental Consequences of Tourism	53
	4.4 Tourism and Lifestyle	55

4.5   Alternative Tourism—Soft Tourism as a Solution?   56
4.6   Ecotourism   58
4.7   Away from Gigantism—But Not All Have Noticed it yet   61
4.8   Community Based Tourism   61
4.9   Pro-Poor-Tourism.   62
4.10   Corona as the Cause of the Ecological Recovery of Tourist
       Destinations   63
4.11   Increasing Tourism or Travel Costs?   63
4.12   Couchsurfing as a Cheap Accommodation Alternative?   64
References.   65

5   Tourism and Infrastructure   69
5.1   Event Tourism   69
5.2   Overtourism   71
5.3   Urban Living and City Tourism   78
5.4   Covid-19's Impact on Tourism Infrastructure   79
References.   80

6   Everyone Wants to Travel—(Almost) No One Wants the Tourists   85
6.1   Tourism, Security and Political Stability   86
6.2   What Tourism Policy?   86
6.3   Crisis Management in Tourism   87
6.4   Tourism as a Search for Meaning—A New Market Niche?   88
6.5   Volunteer Tourism   89
6.6   Global Code of Ethics for Tourism   90
6.7   Travel in the Corona Era   91
References.   92

7   Anti-Tourism Movements.   95
References.   96

8   The Demand for Mobility.   97
8.1   Mobility, Tourism and Spaces   98
8.2   Tourism as the Production of Resonance.   99
8.3   Covid-19 and Mobility.   101
8.4   Future Developments   103
References.   105

9   Right to Mobility?   107
9.1   Freedom and Mobility   107
9.2   Costs of Mobility   108
9.3   Another Mobility Behavior?   109
9.4   Changed Mobility Behavior due to Covid-19?   111
References.   112

10 Solution Approaches and New Ideas 113
    10.1   Sustainable Tourism 114
    10.2   Unification of Environmental Labels 126
    10.3   Slow Tourism 127
    10.4   Resilience in Tourism 130
    10.5   Tourism Awareness 130
    10.6   Shift to More Ecological Transport 131
    10.7   Travel or Tourism Quotas? 132
    10.8   Increasing the Cost of Travel? 133
    10.9   Inclusion of Externalized Costs 133
    10.10   Worldwide Tourism Catalog as a Prerequisite for
             Controlling Tourism 134
    10.11   New Energy Sources in Transport? 134
    10.12   New Living Arrangements 135
    10.13   Creating New, Replicated Originals as an Alternative? 136
    10.14   Space Flights as New Tourism? 136
    References 137
11 Conclusion and Outlook 141

List of Figures

Fig. 2.1   Distinction between tourist and non-tourist mobility. *Source* Scott et al. (2012, p. 3); own translation   4

Fig. 2.2   Tourism volume in millions of arrivals. *Source* Statista (2021b), own representation   7

Fig. 2.3   International tourist arrivals in 2017. *Source* Rosdorff (2020, p. 51) and UNWTO (2018, p. 2); own representation   8

Fig. 2.4   Tourism revenues by continent and region in 2017. *Source* UNWTO (2018, p. 2); own representation   8

Fig. 2.5   Market development in the hotel sector. *Source* Statista (2021a), own representation   9

Fig. 2.6   Tourism as part of and continuation of the leisure industry. *Source* Page and Connell (2014, p. 7); own translation   9

Fig. 2.7   Life cycle of tourist offers. (Own illustration)   11

Fig. 2.8   BCG portfolio concept. *Source* Helm (2009, p. 230); own representation   11

Fig. 2.9   Lifecycle of tourist destinations and their effects on tourism policy. *Source* Gross M. (2017, p. 138); own representation   12

Fig. 2.10   Some important aspects for moments of happiness experienced by tourists. *Source* Schnorbus and Wachowiak (2020, pp. 34 ff.); own summary and representation   15

Fig. 2.11   Africa associations and interest in an Africa trip. *Source* Rosdorff (2020, p. 71); own representation   16

Fig. 2.12   Alternatives for destination decision-making. *Source* Mundt (2013, p. 151); own representation   17

Fig. 2.13   Number of overnight stays in Switzerland by domestic and foreign guests until 2020. *Source* Nussbaumer (2021, p. 19); own representation   22

Fig. 3.1   Tourism system. *Source* Newsome et al. (2013, p. 11), slightly modified and translated from English by the author   28

Fig. 3.2   The three pillars of an economy of tourism. (Mod. after
            Freyer 2011, p. 51; slightly modified and simplified representation
            by the author).   31
Fig. 3.3   Suppliers, markets and consumers in tourism. (Mod. after
            Freyer 2011, p. 307, slightly simplified and modified by
            the author)   32
Fig. 3.4   Cruise passengers per year. (From Jordan 2021, p. 15; own
            representation. ^{a} For the year 2020, the estimate from Wikipedia
            was used due to lack of accurate numbers)   33
Fig. 3.5   Tourism revenue in Spain, France, Italy and Germany. (From
            Schmutz et al. 2020, p. 18; own representation)   34
Fig. 3.6   Product components in tourism. (Mod. after Gross 2017,
            p. 77; own representation)   34
Fig. 3.7   Taxes, fees and charges to national and local authorities.
            (From Gross 2017, p. 146; own representation)   35
Fig. 3.8   Service chain in tourism. (Mod. after Dorsch 2016, p. 35;
            Müller 2011, p. 65; FIF 1995; slightly edited by the author,
            own representation)   38
Fig. 3.9   Tourism employment (VZA) in Switzerland in mountain regions
            and larger cities in 2015. (From Zenhäusern and Kadelbach
            2018, p. 11; own representation)   41
Fig. 4.1   Sustainability types in the booking behavior of tourists in
            Germany in 2011. (From Stettler and Wehrli 2013, p. 169;
            own representation)   55
Fig. 4.2   Sustainability types in the booking behavior of tourists in
            Switzerland in 2011. *Source* Stettler and Wehrli (2013, p. 169);
            own representation.   56
Fig. 4.3   Continuum of possible changes in the global world order and
            in tourism. ^{a}Triple Bottom Line is an accounting system that, in
            contrast to traditional accounting, not only calculates profit and
            loss, but also social and ecological aspects. (From
            Higgins-Desbiolles 2016, p. 226; Translation from
            English by the author)   57
Fig. 4.4   "Magic triangle" of sustainable development and soft tourism.
            (Mod. after Kirstges 2020, p. 142; own representation and slightly
            edited by the author)   58

Fig. 4.5   Different orientation of different ecological and sustainable
            tourism concepts. (From Siegrist et al. 2015, p. 20; Siegrist
            and Ketterer Bonnelame 2016, p. 48; own representation)   60

Fig. 5.1   Events at a glance. (Mod. after Freyer 2000, p. 352; Müller
            2011, p. 61; own representation and slightly edited by the author).   70

Fig. 5.2   The attitude of the people of Barcelona towards tourist
            activities in their city. (From Martín et al. 2018, p. 8;
            own representation)   73

Fig. 5.3   Hotspots for tourist concentrations according to a survey
            in Amsterdam. (From Gerritsma 2019, p. 138; own representation). . . .   75

Fig. 5.4   Number of tourist overnight stays per inhabitant in 2018.
            Source Weber (2019, p. 9); own representation.   76

Fig. 6.1   Peace—intercultural understanding security cycle.
            (Mod. after Edgell and Swanson 2013, p. 12; own representation).   86

Fig. 8.1   Mutually reinforcing technical, social and everyday acceleration.
            (Mod. after Rosa 2013, p. 44; own representation)   98

Fig. 8.2   Resonance spheres, experience spaces and suitable travel
            forms in tourism. (Mod. after Aschauer 2020, p. 59; own
            representation)   100

Fig. 8.3   The growing number of tourists correlates with an
            increasingly negative attitude of the local population.
            (Mod. after Coghlan 2019, p. 73; own representation)   105

Fig. 9.1   Cost distribution for motorized road traffic in Switzerland, a
            total of 72 billion Swiss francs in 2015. (Mod. after Quandt
            and Gigon 2019, p. 43; own representation)   109

Fig. 9.2   Cost distribution for rail traffic in Switzerland, a total of
            11 billion Swiss francs in 2015. (Mod. after Quandt and
            Gigon 2019, p. 43; own representation).   110

Fig. 10.1   Holiday effects. (Mod. after Lohmann 2019, p. 18, own
            representation, slightly supplemented by the author)   114

Fig. 10.2   Milestones in a comprehensive concept of sustainable
            development. (Mod. after Reddy and Wilkes 2013, p. 6
            and own research; own representation)   115

Fig. 10.3   Manifest and latent network relationships in tourist clusters.
            (Mod. nach Schuhbert 2018, p. 244; eigene Darstellung).   124

Fig. 10.4   Motivation to go hiking in the different age groups.
            (Mod. after Dreyer and Dürkop 2011, p. 107; own representation)   128

Fig. 10.5   Slow Travel as a holistic experience of travel.
            (From Dickinson and Lumsdon 2013, p. 375).   129

List of Tables

Table 2.1   Forms of tourism   6
Table 2.2   Individual holiday types by frequency   8
Table 3.1   Destination types.   28
Table 3.2   Determinants for tourist spending behavior.   29
Table 3.3   Development of elasticity in demand for overnight stays
            in international travel between 1960 and 1985 in relation
            to gross domestic product and real income.   29
Table 3.4   The world's largest hotel chains measured by the number
            of rooms.   33
Table 3.5   Eight advantages and eight disadvantages of tourism.   36
Table 3.6   Cost structure of an independent travel agency.   43
Table 4.1 CO_2 emissions by type of holiday.   48
Table 4.2   Possible adaptation measures to direct and indirect
            consequences of climate change in Germany.   54
Table 4.3   Advantages and disadvantages of couch surfing from
            the perspective of the guest.   64
Table 5.1   Development of some key figures for the city and tourism
            in Venice between 2007 and 2017.   74
Table 5.2   Visitor to resident ratio in 2018 in Lucerne and Venice.   77
Table 6.1   External, internal, exogenous and endogenous causes of crises.   88
Table 6.2   Material and immaterial effects of crises on companies
            or industries.   88
Table 6.3   Forms of engagement in volunteer tourism.   90
Table 8.1   Different types of mobility.   98
Table 8.2   Decline in air traffic 2020.   102
Table 8.3   Opportunities and risks for tourism in the next 10–20 years.   104
Table 9.1   Transportation costs borne by the general public in Switzerland.   110
Table 9.2   Cost comparison of car purchase and car rental.   110
Table 10.1 Core indicators for sustainable tourism management.   118
Table 10.2 Core economic indicators.   120

Table 10.3   Core indicators for the socio-cultural environment.   121
Table 10.4   Core indicators for the ecological dimension of tourism.   122
Table 10.5   Environmental labels for hotels in Switzerland, as of June 2017.   126

Introduction

According to Hans Magnus Enzensberger (1962, p. 152), "traveling belongs to the oldest and most widespread figures of human life". Enzensberger (1962, p. 156) saw the roots of tourism in English, French and German Romanticism. The escape from the self-created reality was made possible by the very means of communication with which this new, modern reality had been created.

In contrast to this rather skeptical view of tourism, at the end of the twentieth century the euphoric views of tourism were overwhelming. For example, Przeclawski (2016, p. 126) wrote: "At the end of the twentieth century, humanity enters a new phase—the phase of the touristification of the world. This process mainly affects developed countries, but gradually spreads all over the world. Tourism becomes a 'way of life' of today's human being."

Spode (2020, p. 19) has pointed out that since the 1960s, it has been possible to speak of "two branches of tourism research", on the one hand of a classical and relatively homogeneous research strand, which is oriented towards questions of economic and political steering of tourism and which is mainly concerned with "how-questions", and on the other hand of a more heterogeneous research tradition, which is concerned with social analyses and "why-questions". There was hardly any contact between the two and the intersections were too small for there to be a real cooperation between the two sides (Spode 2020, p. 19). Surprisingly, in recent years, the importance of economic sciences has decreased and that of the humanities has increased in tourism-related (basic) research papers and dissertations in the English-speaking world, above all through psychological, but also multidisciplinary work (2020, p. 20). However, this view is somewhat one-sided because a large part of the scientific work on tourism is created at universities of applied sciences, which hardly have doctoral students and which carry out little basic research, but much more applied research. Accordingly, Spode (2020, p. 22) believes that there will not be an actual or integrating "theory of tourism". Rather, it is about interactions, about a interplay of social and economic factors and about applications in a research field that can be described as "tourism". For the same reason, tourism is often also seen as a cross-sectional topic with a strong transdisciplinary orientation.

Since the 1990s, the so-called *Mobility Studies* have developed more and more into an interdisciplinary field of research. This led to a veritable *"mobility turn"* (Urry 2007, p. 6). The associated new view of mobility linked the analysis of various forms of travel, transport and communication with economic and social life through different time references and spaces. According to Schiele (2017, p. 8), the *"mobility turn"* starts from spatial analyses, but takes a perspective change, according to which space is primarily socially structured, but mobile. Seen in this way, travel and tourism mean the production and maintenance of spatial and temporal relationships—they thus shape a kind of social space. Questions of identity, self- and other-view, belonging and demarcation also become important here.

Kerstin Heuwinkel (2019, p. 11) has pointed out in the introduction to her "Tourism Sociology" that in Germany in the literature on tourism, economic and geographical approaches predominate, while in other European countries cultural aspects are also addressed. Heuwinkel (2019, pp. 11 ff.) deals with tourism as a social phenomenon. This is because so far there has been a lack of systematic analysis on the basis of sociological theories and methods. But tourism also has an important economic function. This has become particularly clear in the Corona pandemic.

References

Enzensberger, Hans Magnus 1962: Eine Theorie des Tourismus. In: Enzensberger, Hans Magnus: Einzelheiten. Frankfurt/Main: Suhrkamp. 147 ff.
Heuwinkel, Kerstin 2019: Tourismussoziologie. München: UVK.
Przeclawski, Krzysztof 2016: Deontology of Tourism. In: Fennell, David (Hrsg.): Tourism Ethics. Critical Concepts in Tourism. Volume 1: Theories of Ethics and Tourism. London/New York: Routledge. 117 ff. Ursprünglich in: Progress in Tourism and Hospitality Research. 2 (1996). 236 ff.
Schiele, Kertin 2017: Tourismus und Identität. Vietnam-Reisen als Identitätsarbeit von in Deutschland lebenden Việt Kiều. Berlin: regiospectra.
Spode, Hasso 2020: Tourismus und Gesellschaft: eine neualte Forschungsaufgabe. In: Reif, Julian/ Eisenstein, Bernd (Hrsg.): Tourismus und Gesellschaft. Kontakte – Konflikte – Konzepte. Schriften zu Tourismus und Freizeit. Band 24. Berlin: Erich Schmidt Verlag. 17 ff.
Urry, John 2007: Mobilities. Cambridge/UK/Malden/MA: Polity Press

Tourism—A Short History

2

In the sense understood today, tourism is a newer phenomenon. But at all times and in all places there have been travelers—often only in small numbers.

2.1 To the Term

Starting from the question of what a tourist or a tourist would be, Heuwinkel (2019, p. 14) has pointed out that there is no verb for the activity of the tourist. A traveler travels, a migrant migrates or a refugee flees. It is interesting that many people who are on the road or traveling do not want to be tourists—they are travelers, researchers, people on the move or nomads. Obviously, the term "tourist" has acquired a negative connotation—similar to how "asylum seeker" or "foreigner" has developed over time.

In Veste's opinion (1999, p. 50), tourism can be "regarded as a large-scale attempt to organize emotions. Pleasure, fun ('fun'), cheerfulness, gaiety, exuberance, euphoria, bliss are among the emotional experiences that one expects from travel"—and all this is more or less promised by the tourism industry, without however There are no guarantees for emotions or there can be (cf. also Steinbach 2003, p. 112).

According to Freyer (2015, p. 3), the *change of location*, that is, the movement from the normal place of residence to a "foreign place", the *stay* in a foreign place and the *motives* why a change of location is made, why is traveled.

Figure 2.1 sketches a possible distinction between tourist and non-tourist mobility.

The World Tourism Organization UNWTO has proposed the following definitions and delimitations:

- A traveller is someone who moves between two geographical locations, regardless of the purpose or duration.

C. J. Jäggi, *Tourism Before, During and After Corona*, https://doi.org/10.1007/978-3-658-39182-9_2

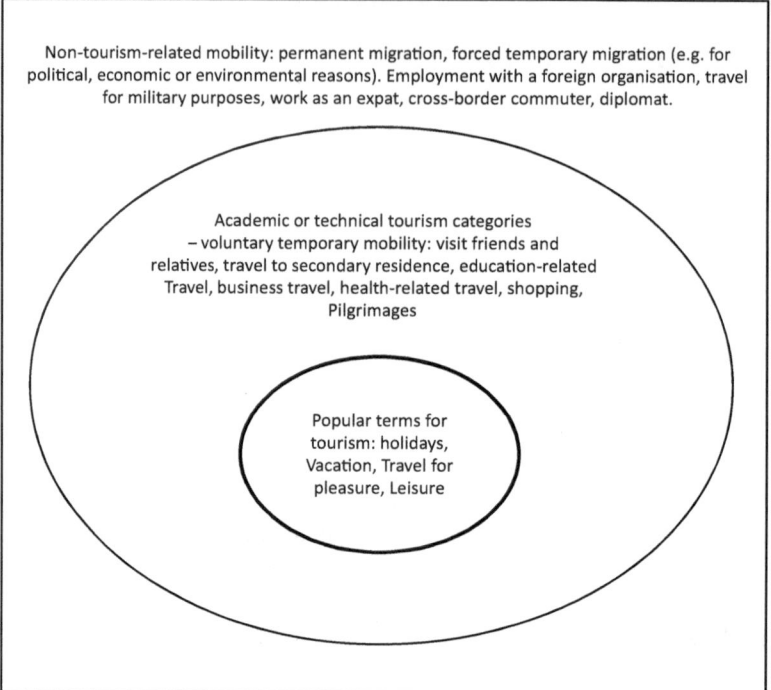

Fig. 2.1 Distinction between tourist and non-tourist mobility. *Source* Scott et al. (2012, p. 3); own translation

- A visitor is a traveller who makes a journey to a destination outside his or her usual environment, for less than a year, for business purposes, as a leisure activity or for personal reasons, but without being employed in the destination country or at the destination.
- A visitor is classified as a tourist or an overnight visitor if his or her trip includes at least one overnight stay or if he or she returns as a day visitor or excursionist on the same day (Graham and Dobruszkes 2019, p. 3).

In contrast, Graham and Dobruszkes (2019, p. 3) have proposed using the term tourism more broadly: tourist services as satisfaction of both personal and business demand.

For a long time, the term "Fremdenverkehr" was used in German-speaking countries[1]. Today, the term "tourism" is also used in German-speaking countries. In some German-speaking countries, "tourism" was used for domestic tourism ("domestic tourism"), while "tourism" was used for interstate travel (Gross S. 2017, S. 31). However, the term

---

[1] The German word ‚Fremdenverkehr' [comprises the] totality of relations and phenomena resulting from the change of place and the stay of persons for whom the place of stay is neither the principal and permanent place of residence nor the place of abode" (Kaspar 1996:16, cited in Schiele 2017:59). Today German speakingcountries talk also about "tourism".

"tourism" is rarely used in tourism research because the word "foreign" should not be in the center, but rather "guest" and "hospitality". In addition, the term "tourism" has established itself internationally (Gross S. 2017, S. 32).

Schiele (2017, S. 59) has pointed out that the many phenomena and tourist services that are related to each other give tourism a system character.

According to the UNWTO, individual, non-business demand for tourist services can be divided into the following categories: 1) holidays, leisure and recreation, 2) visits to friends and relatives, 3) education and training, 4) health and medical care, 5) religious travel and pilgrimages, 6) shopping, 7) transit area and 8) others (Graham and Dobruszkes 2019, S. 3). According to the UNWTO, in 2017, tourism accounted for 13% of business travel, 53% of leisure, recreation and leisure travel, and 27% of visits to friends and relatives, as well as travel for health, religious or other reasons. 7% did not give any information (Graham and Dobruszkes 2019, S. 4).

Since the 1970s, four basic criteria or elements have been distinguished for tourism: 1) the travel motive, 2) the mode of transport, 3) the length of stay and 4) the distance traveled (Gross S. 2017, S. 35).

Table 2.1 lists the various forms of tourism according to different categories such as location, duration, purpose, etc.

Based on this, the following areas of the tourism industry can be distinguished: accommodation; restaurants; rail transport, road transport, water transport or air transport of passengers; transport of equipment and luggage, travel agencies and reservation service activities, cultural activities, sports and recreation activities, retail of country-specific offers for tourists, other country-specific tourism activities (Graham and Dobruszkes 2019, p. 4).

2.2   Some Numbers and Facts

Since 1950, global tourism has been growing continuously. Up to the Corona pandemic, tourism was one of the fastest growing economic sectors (Hartmann 2020, p. V). In particular, since 1985, the numbers have exploded. Figure 2.2 shows the development of global arrivals until 2019.

In 2017, the following five markets were responsible for almost 60% of global online travel bookings in this order: the USA, China, Japan, Germany and the United Kingdom (Hanke 2019, p. 108).

According to Christine Plüss (2019, p. 55), managing director of the Tourism and Development Working Group, as of 2019 only a small minority of the worl's population was able to travel internationally. So by 2019, less than 10% of the world's population had ever boarded an airplane. If this is true and one takes into account the growth of the middle class in populous countries such as China, India, Indonesia, Nigeria or Pakistan, the potential for tourism is enormous—but so is the danger of increasing overtourism [2].

---

[2] For a detailed discussion of the question of overtourism, see Sect. 5.2 "Overtourism".

Table 2.1 Forms of tourism

Distinction by							
Motivation							
Motiv	Business	Health	VFR[a]	Religion	Other	Recreation, holiday	Study, Work, Emigrate
Designation	Business tourism	Health tourism	VFR[a]-Tourism	Religious tourism	E.g. cultural, sports tourism	Recreational, vacation tourism	Working, study visits, emigration
Duration							
Days	1	1–4			5–30	Over 30	Over 1 year
Overnight stays	0	1–3			4–29	Up to 1 year	
Designation	Day trip	Short-term Tourism			Recreational tourism	Long-term tourism	Extended stay
Destination							
Distance	(Home)	Closer surroundings			Inland	Abroad	To work
Name	Location					Continent	Minor border traffic
	City tourism	Near tourism/recreation			Domestic tourism	Foreign tourism	(Professional) commuters

Source: Groß S. (2017, p. 37); own representation.
[a]Visits of friends and relatives

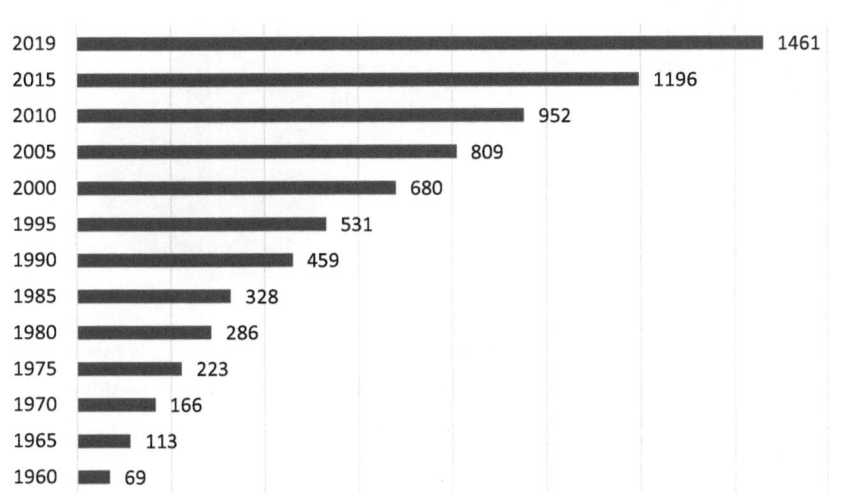

Worldwide tourism volume in millions travel arrivals

Fig. 2.2 Tourism volume in millions of arrivals. *Source* Statista (2021b), own representation

Tourism is still very unevenly distributed over the individual continents. For example, before the Corona pandemic, the continent of Africa had only 5% of global tourist arrivals—and only 3% of tourism revenues (Hartmann 2020, p. V; UNWTO 2018, p. 2). In contrast, 51% of international tourists traveled to Europe in 2017 (UNWTO 2018, p. 2). Figure 2.3 shows the distribution of international tourist arrivals by continent and region.

In contrast, tourism revenues were distributed very differently. In 2017, only 39% of revenues went to Europe, 29% to Asia and the Pacific, and 24% to the American continent (UNWTO 2018, p. 2). Figure 2.4 shows tourism revenues for the year 2017 by continent and region.

Many different types of holidays are included under tourism. Table 2.2 shows the distribution among the individual holiday types. However, it should be noted that the different holiday types can also overlap.

Accommodation and hospitality form an important part of the tourism industry. In 2018, the global hotel industry generated revenues of around US$600 billion. Figure 2.5 shows the development of the global hotel industry between 2014 and 2018.

The hospitality industry in the narrower sense generated US$554 billion worldwide in 2018.

Before the Corona year 2020, China was the country with the world's largest tourism expenditure, followed by the USA and Germany. In 2018, global tourism revenue amounted to US$1,451 billion (Statista 2021c).

Fig. 2.3 International tourist arrivals in 2017. *Source* Rosdorff (2020, p. 51) and UNWTO (2018, p. 2); own representation

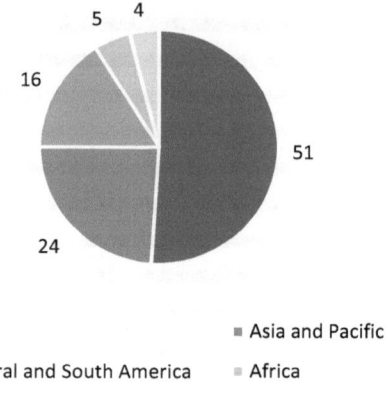

International tourism arrivals in % in 2017

- Europe
- Asia and Pacific
- North, Central and South America
- Africa
- Middle East

Fig. 2.4 Tourism revenues by continent and region in 2017. *Source* UNWTO (2018, p. 2); own representation

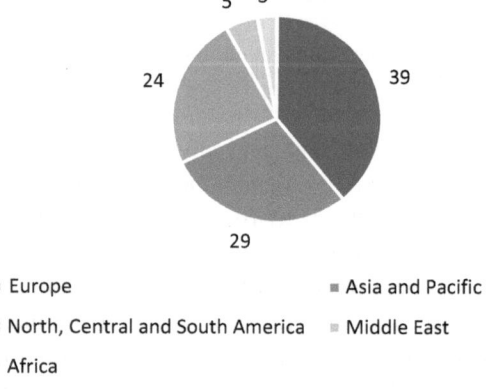

Revenue from tourism 2017 by continent and Region in %

- Europe
- Asia and Pacific
- North, Central and South America
- Middle East
- Africa

Table 2.2 Individual holiday types by frequency

Type of vacation of Germans (Multiple answers possible)	In percent of vacation trips
Beach/bathing/sun vacation	44%
Rest vacation	35%
Nature vacation	27%
Active vacation	16%
Health vacation*	6%

Source: Gross M. (2017, p. 10)

*However, health travel and stays in Germany are subject to an atypical demand situation because in Germany the health system requires a duration of at least three weeks (21 days) by law. No private vacation days have to be used. Source: Gross M. 2017, p. 10).

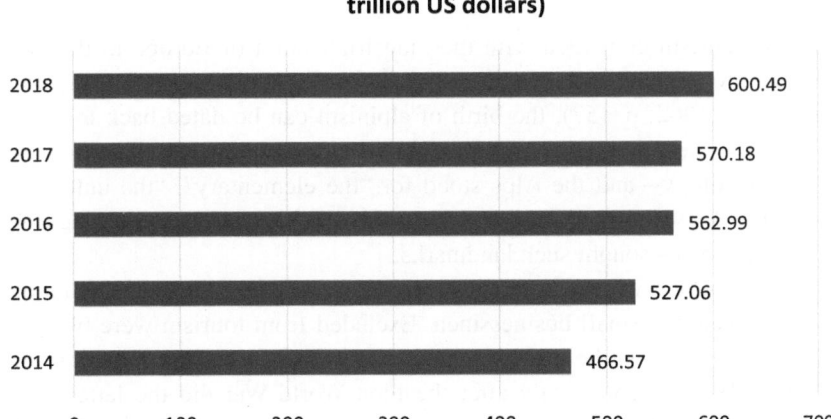

Fig. 2.5 Market development in the hotel sector. *Source* Statista (2021a), own representation

In addition, there is something on the planet called "global urbanization": While in 1950 only 30% of the world's population lived in cities, in 2014 it was 54%, and by 2050 it is estimated to be 66% (Rate et al. 2018, p. 3).

From another perspective, tourism can also be seen as part of or continuation of the leisure industry (Fig. 2.6).

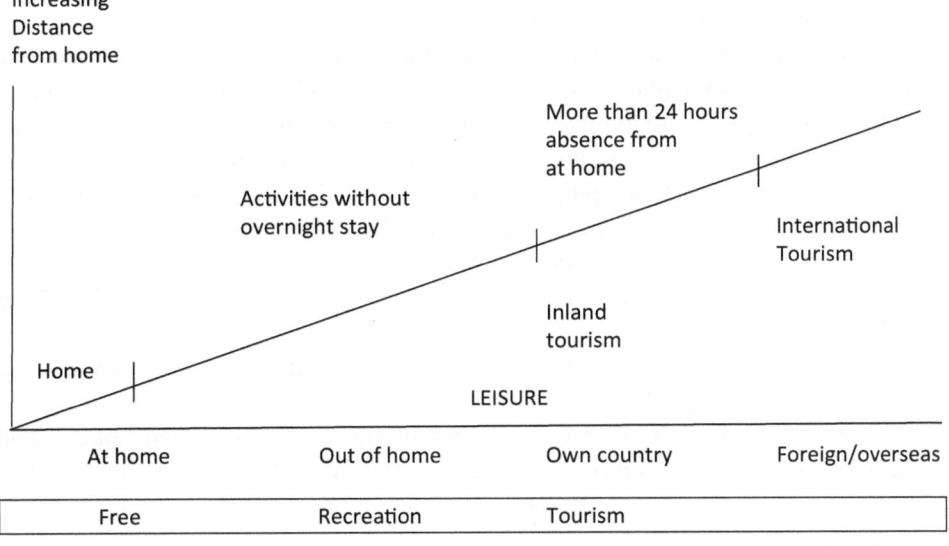

Fig. 2.6 Tourism as part of and continuation of the leisure industry. *Source* Page and Connell (2014, p. 7); own translation

2.3   Development

The emerging tourism coincided with the industrialization of Europe in the nineteenth century. A key role in the spreading tourism played the mountaineers. Thus, according to Enzensberger (1962, p. 157), the birth of alpinism can be dated back to 1787, when Saussure became the first to climb Mont Blanc. The expedition into the Alps embodied the romantic ideology—and the Alps stood for "the elementary", "the untouched" and "the adventure" (Enzensberger 1962, p. 157). Even today, tourists—according to Enzensberger (1962, p. 159)—sought such landmarks.

First of all, members of the upper class traveled, later all members of the middle class, officials, craftsmen and small businessmen. Excluded from tourism were two groups of people for a long time: on the one hand the peasants and on the other hand the workers (Enzensberger 1962, p. 160). Only after the First World War did the latter receive the opportunity to escape the—as Enzensberger (1962, p. 160) says—"enormous pressure of the industrial working world at least for the duration of a few weeks, at least apparently". New collective agreements after the First World War contained clauses for paid holidays for the first time. As late as 1940, only about a quarter of American workers were entitled to paid leave. By 1957, this figure had risen to 90% (Enzensberger 1962, p. 160). With the widespread access to vacation travel, the attitude towards foreign people and countries also changed. Quickly, sightseeing turned into "life-seeing" (Enzensberger 1962, p. 164), the lifestyles and customs of other countries of the ethnic groups became a tourist event.

In the process, tourist offers also developed and changed. Like any product and any service, vacation offers, destinations and tourist trends go through life cycles. Figure 2.7 shows the emergence, development, peak and decline of tourist offers.

However, both the initial position, expansion and peak as well as the final state can vary greatly. This also applies to new and old tourism and leisure offers, destinations, travel forms, perhaps also to tourism in a certain region as a whole. Determinants for the demand [3] and their development are, according to Graham and Metz (2019, p. 43), the average number of trips and the proportion of the population that undertakes trips. What is particularly characteristic of tourist life cycles is that often—but not always—the transition to mass and cheap tourism can also be a sign of the decline of the relevant offer, even if it may still appear to be "Cash Cow" in terms of the Boston Consulting Group portfolio.

The portfolio analysis of the Boston Consulting Group (BCG) shows which development cycle a product or product line goes through (Fig. 2.8).

The most interesting from an economic point of view are the so-called *"stars"* (high market growth and high market share), the least interesting and shortly before the end the

---

[3] The two authors do refer to air travel, but the same determinants apply by analogy to the other tourism offers.

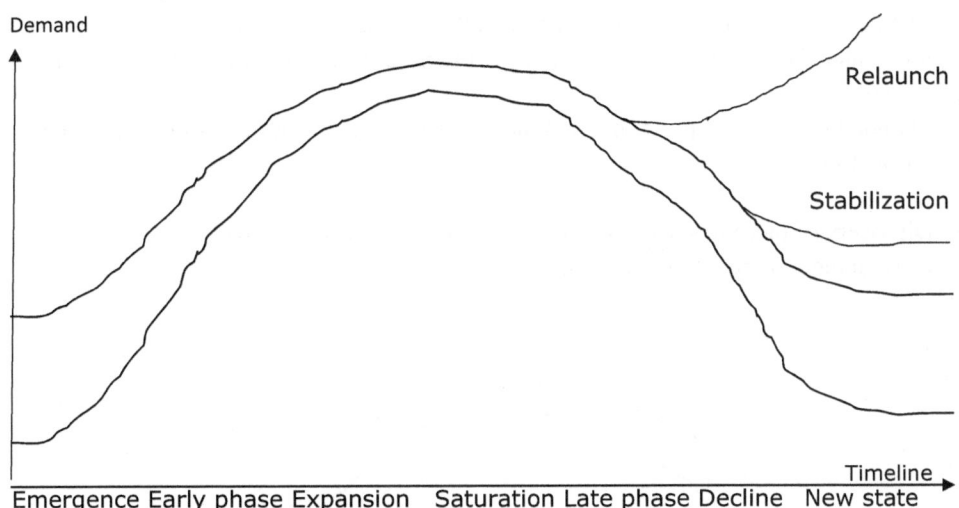

Fig. 2.7 Life cycle of tourist offers. (Own illustration)

>>1	▲	Question Marks	Stars
Relative market growth		Introductory Phase	Growth phase
		poor dogs	**cash cows**
<<1		Degeneration phase	Maturity phase
		<<1 **Relative market share** >> 1	

Fig. 2.8 BCG portfolio concept. *Source* Helm (2009, p. 230); own representation

"poor dogs" with low market growth and small market share. In contrast, *"cash cows"* are economically interesting in the short term, but it is foreseeable when their potential is exhausted. The intention of corporate strategy should be to make *"question marks"* into *"stars"* and to keep them there. If the market potential then sinks, the products are milked as *"cash cows"* as long as possible. In the end, they become *"poor dogs"* and finally disappear. *"Stars"* need secure financing, but also generate a lot of income, while *"cash cows"* have exhausted their potential at some point and can then only be maintained with excessive financial resources. They then become *"poor dogs"* and should then be given up as a business unit. The portfolio should be as balanced as possible, with a sufficiently large number of *"cash cows"* that should generate 40-60% of total sales (Domschke and Scholl 2008, p. 185).

Matilde Gross (2017, p. 138) has graphically represented the life cycle of tourist destinations and the effects on tourism policy. Figure 2.9 transfers this relationship to the schema of Fig. 2.7.

Matilde Gross (2017, p. 42) has summarized the life cycle of thermal spas and seaside resorts as follows:

- Discovery of thermal springs and establishment of seaside resorts,
- Development of health treatments,

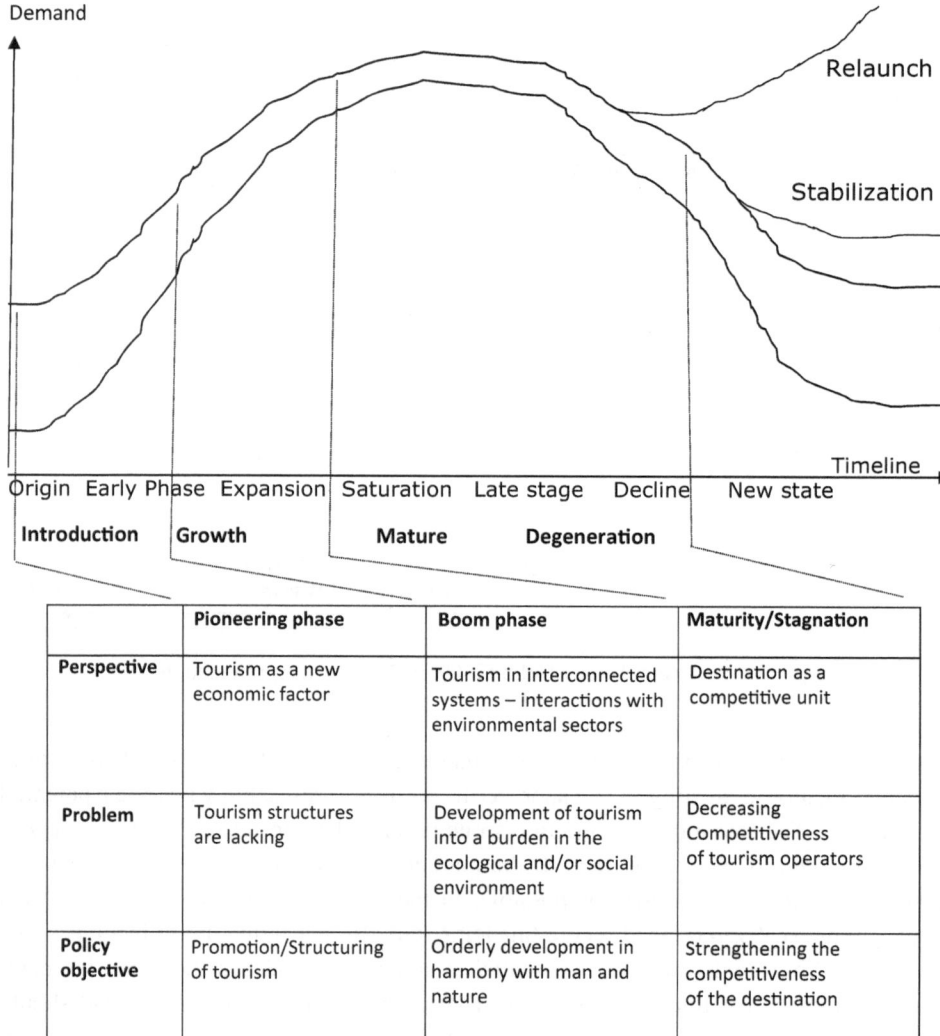

	Pioneering phase	Boom phase	Maturity/Stagnation
Perspective	Tourism as a new economic factor	Tourism in interconnected systems – interactions with environmental sectors	Destination as a competitive unit
Problem	Tourism structures are lacking	Development of tourism into a burden in the ecological and/or social environment	Decreasing Competitiveness of tourism operators
Policy objective	Promotion/Structuring of tourism	Orderly development in harmony with man and nature	Strengthening the competitiveness of the destination

Fig. 2.9 Lifecycle of tourist destinations and their effects on tourism policy. *Source* Gross M. (2017, p. 138); own representation

- Increasing importance and peak of popularity in the eighteenth and nineteenth centuries,
- Growing insignificance after the two world wars,
- Increasing deterioration of infrastructure towards the end of the twentieth century, e.g. due to changed financial conditions,
- Musty image, loss of attractiveness and marketing problems of many spa towns and health resorts as well as
- Partial new boom in the twenty-first century through management-oriented revitalization measures.

Trendy tourism offers are very much dependent on fashion and zeitgeist. Kirstges (2020, p. 22) even speaks of a "growing obsolescence of tourism products" with regard to many travel destinations, i.e. of a "aging" or "over-aging" of the travel offer.

Steinbach (2003, pp. 246 ff.) has also observed a product and maturity cycle in holiday styles: an initial innovation phase by so-called "trendsetters", that is, the first individual demanders, mostly from affluent and financially strong circles, followed by a growth phase in which the holiday style is adopted by a larger number of demanders, but usually not with the same financial resources. The travel organizers include the type of holiday in their program. After a more or less long expansion phase, in which a development towards an "introverted" mass tourism for guests with reduced financial means takes place, a maturity and shrinkage phase follows, in which a so-called "buyer's market" develops towards a growing share of cheap and last-minute offers—and the providers begin to take the increasingly less attractive travel, destinations and holiday forms out of their travel offer (Steinbach 2003, p. 247). If this is true, there is a significant "use-up or burn-out effect" of tourism offers, but also of destinations, and these have to be spruced up, renewed or rebuilt. All of this is hardly sustainable and usually associated with an enormous demand for resources, namely in infrastructure, marketing and in terms of the environment. The socio-cultural context is also increasingly emptied and degenerates into pure folklore.

With a view to the differently developed countries, Müller (2011, p. 29) has formulated various tourist challenges for the individual countries:

- Less developed countries: Removal of market entry barriers, lower transport costs, simplification of visa requirements, loosening or abolition of import quotas, reduction of foreign influence.
- Emerging countries: Transformation of rapid growth into sustainable growth, expansion of infrastructure, harmonization of capacities and removal of bottlenecks.
- Highly developed countries: Overcoming the weakness of growth, strengthening competitiveness in terms of the production factors of labor and capital, overcoming fragmented structures.

Leaving aside the fact that all of these factors are based on an unreflective, purely quantitative understanding of growth in the sense of "the more—the better" and that today, especially in tourism, an orientation towards a qualitatively and differentially understood growth is required (Jäggi 2021, pp. 119 ff.), this model ignores the following: The transformation of society from an economy based on fossil energy sources to a sustainable society is urgently needed in tourism in particular. Here, sustainable development models are required that do not shy away from completely new concepts—not even from a partial restriction, slowdown or even reversal of growth *(Degrowth)*.

2.4   Reasons to Travel

Rate et al. (2018, p. 4) have rightly pointed to the ever-increasing diversity of people's lifestyles, world views and living conditions. This is not only reflected in the growing heterogeneity of travel motivations—it also makes it increasingly difficult to develop standardized travel offers for everyone. Accordingly, mass tourism is not only quantitatively, but also qualitatively increasingly reaching its limits. Rate et al. (2018, p. 5) therefore believe that more and more travelers are looking for holistic models of recreation, that is, offers that address body, soul and mind. Instead of "travel consumers" ("consumers") there are more and more "transumers" ("transumers"), who are looking for depth and personal development instead of diversity and distraction. At the same time, tourists are increasingly focusing on digital offers and services. Today there are almost endless possibilities for the combination, integration and synthesis of marketing, technology and digital media. A new type of tourist is the so-called "prosumer" ("prosumer"; Rate et al. 2018, p. 8), who is actively involved in the design of the travel product—interactively, proactively and reactively.

In the sense of Cohen (1979), there are the two dimensions of *tourist perception* and the *type of offer,* where the tourist background can be either real or staged and tourists can perceive the experience space or the offer as either authentic or artificial or staged (Heuwinkel 2019, p. 56). According to Cohen (1979), there are mainly five types of tourist experience and thus also motivation for tourist travel to be distinguished: 1) relaxation or recreation *("recreation"),* 2) variety or distraction *("diversion"),* 3) experience *("experience"),* 4) experiment *("experiment")* and 5) experience of being *("existence";* Heuwinkel 2019, p. 58). However, these motivations can also flow into each other.

Central to tourism are the emotions associated with a trip or vacation—these include anticipation, excitement, curiosity, joy and a sense of adventure, but also fears, for example in the face of uncertainties, or shock experiences, for example when confronted directly with abject living conditions (Heuwinkel 2019, p. 164).

Schnorbus and Wachowiak (2020, pp. 34ff) emphasized above all the importance of physical and psychological destination attributes and the significance of distance from everyday life as elements for tourist moments of happiness. Figure 2.10 shows the most important aspects.

Happiness aspects for tourists in %

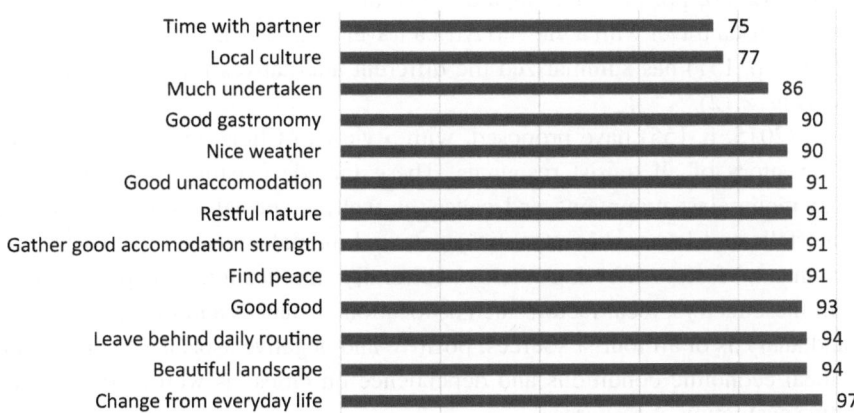

Time with partner	75
Local culture	77
Much undertaken	86
Good gastronomy	90
Nice weather	90
Good unaccomodation	91
Restful nature	91
Gather good accomodation strength	91
Find peace	91
Good food	93
Leave behind daily routine	94
Beautiful landscape	94
Change from everyday life	97

Fig. 2.10 Some important aspects for moments of happiness experienced by tourists. *Source* Schnorbus and Wachowiak (2020, pp. 34 ff.); own summary and representation

Aschauer (2020, p. 52) has compiled six central theory fragments of current research on travel motivation:

- Enzensberger's theory of flight or deficit,
- the theory of saturation, which interprets travel as compensation and contrast to everyday life,
- the theory of authenticity, according to which travel is motivated by the search for originality,
- the self-activation theory, which is based on the basic need for self-development,
- the theory of self-staging as an expression of the achievement society professionally as well as individually, for example in order to achieve social recognition,
- theory of experience orientation through stimulus seeking and search for activities, which should lead to a kind of flow state, from wellness to culinary.

All these theories have in common that travel and vacation are used to seek an interruption or distance from everyday life—associated with a new self-positioning. So seen, travel can also be seen as an attempt at a temporally limited counter-draft to everyday life.

2.5   Travel Destinations

The respective travel destination is closely linked to the travel motivation. The travelers want to experience something, see something or achieve a specific goal—for example in sports tourism, in health tourism or—to a certain extent—in religious travel or pilgrimages.

Both the general image of the destination and the specific motivation for the trip play a role. Rosdorff (2020, pp. 71 ff.) Compared general Africa associations and resulting motivations for Africa travel with a view to Africa travel (Fig. 2.11).

Mundt (2013, p. 151) has summarized the different alternatives for destination decision-making (Fig. 2.12).

Chen et al. (2015, p. 155) have proposed, with a view to China, to carry out a comprehensive inventory of all tourist resources. These include existing rock formations, tectonics and mountains, mountains and caves, as well as special landforms, sources, streams, waterfalls and lakes, characteristic plants and animals, meteorological and climatic environmental factors affecting tourism, culturally significant landscapes, aesthetic aspects of all these factors, local factors such as location, distances and transport options, overview and analysis of all tourist sources, positive and negative aspects of neighboring resources, local economic conditions and dependence on cities, as well as quantitative and qualitative evaluation of resources.

Such a tourism cadastre could in future form the basis for possible contingentiations and travel restrictions—especially if the demand for tourist and travel destinations continues to increase.

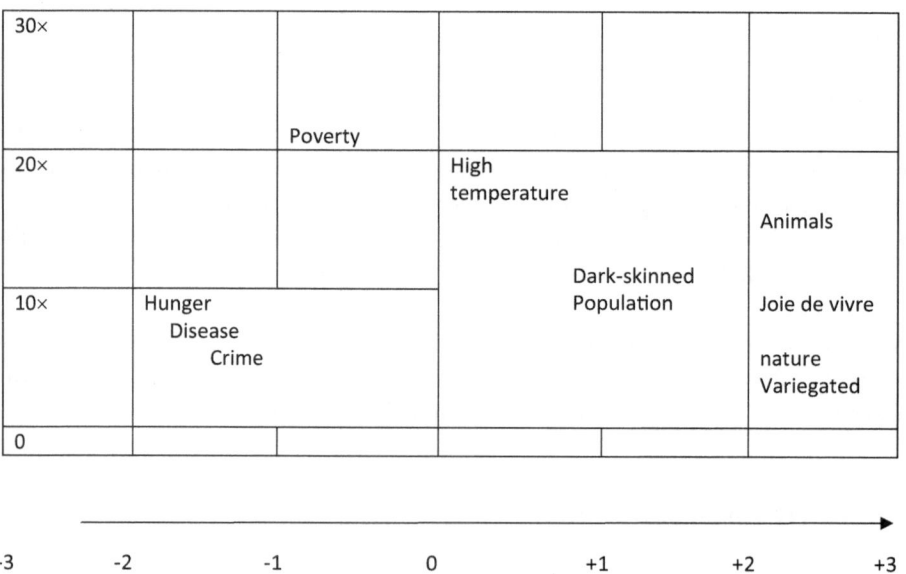

Fig. 2.11 Africa associations and interest in an Africa trip. *Source* Rosdorff (2020, p. 71); own representation

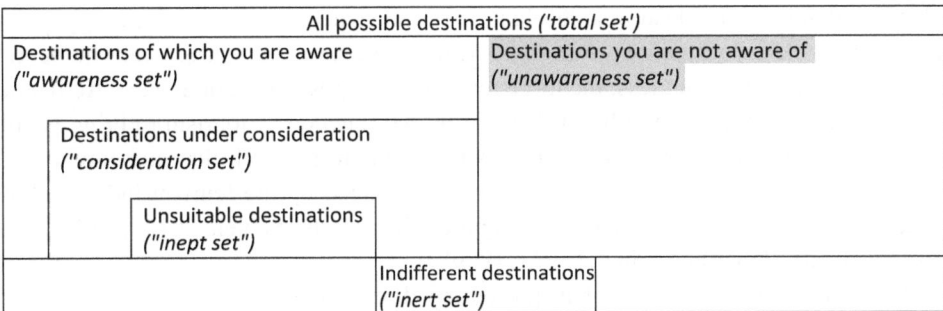

Fig. 2.12 Alternatives for destination decision-making. *Source* Mundt (2013, p. 151); own representation

A study by the Swiss National Science Foundation on the attractiveness of tourist destinations for the population and the relationship between infrastructure, accessibility and tourist demand came to the conclusion in 2011 that the choice of destination is essentially dependent on accessibility (Tschopp et al. 2011, p. 117). Thus, the attractiveness of a visit is rated by travellers to a much greater extent in dependence on the distance and travel time from their place of residence to the possible destinations and compared with possible alternatives. Municipalities located at a great distance from population centres therefore have to offer much better tourist facilities in order to compensate for this disadvantage. Or these municipalities improve their connection to the transport network in order to remain competitive. For example, the presence of a railway station has a positive effect on the probability of a visit and thus on the number of visitors (Tschopp et al. 2011, p. 117). In contrast, the type of equipment with infrastructure facilities for leisure activities has a comparatively small impact on the choice of destination. Although the basic facilities—e.g. ski lifts for skiers—must be available, their quality is of secondary importance. What is more important is the diversity of the infrastructure (Tschopp et al. 2011, p. 117).

2.6 Tourist Gaze

An important concept in tourism studies is the so-called *"tourist gaze"*, that is, the tourist gaze or the gaze regime (Schiele 2017, p. 94). In their work of the same name, Urry and Larsen (2011, p. 2) have defined the tourist gaze as a filter of ideas, skills, wishes and expectations which are predetermined by social class, gender, nationality, age and education. This tourist gaze orders, delineates and evaluates the world to a greater extent than it reflects it.

In this context, the question of the Other and of its authenticity also arises. Schiele (2017, p. 121) was quite right to point out that the search for the very Other often entails a distance that can often lead to a devaluation of the visited—or even to an exaggeration of the exotic and a devaluation of the self. At the same time, the tourist seeks authenticity, which he finds less and less, the more tourism takes possession of a place. Authenticity also means untouchedness; but it is precisely the function of tourism to bring people to untouched places. It has long been known in tourism research that authenticity is a construct that is deliberately instrumentalized and staged by the tourism industry, both in the destination and in the country of origin of the tourists (Schiele 2017, p. 121). This was already noted by MacCannell (1973, 1999).

Many tourist strategies consist in presenting destinations as expressions of uniqueness—or, as Schäfer (2015, p. 28) puts it, as "markers of difference". Tourism sells uniqueness, which, however, becomes increasingly problematic in times of general "crisis of representation", in which the original can hardly be distinguished from the copy. The marking of one's own identity is closely linked to this, which often happens precisely in the emphasis on the difference to the "Other" and thus also becomes the central theme in tourism (Schäfer 2015, p. 28). The comparison of "there" and "with us" is central on all levels: starting with natural sights, in culture, in infrastructure, in food, accommodation, etc. Depending on the country and place, alleged authenticity or—to put it badly—typology is sold: from "Scottishness", via "Swissness" to "Africa" or "Wildlife". The old, the traditional, the heritage, the history are considered "authentic" (Schäfer 2015, p. 29). For example, national parks and nature are turned into "naturalness" (Newsome et al. 2013, p. 4) and well-being or health into "wellness".

The idea is not new, to understand tourist destinations as social constructions. But the changes in the perception of the individual destinations have been little researched so far (Saarinen 2014, p. 56).

2.7   Tourism During and After Corona

The Covid-19 pandemic led to a massive decline or even a complete stop of foreign tourism in a number of countries. For example, Cambodia stopped issuing tourist visas from mid-March 2020 until well into 2021 (Mihai 2021, p. 10). In March 2020, Japan closed its borders to foreign tourists—after Japan had been visited by 32 million foreign tourists in 2019 (Putz 2020, p. 3).

Thailand, which according to official figures kept the Corona virus in check to a very successful extent in 2020 (Peer 2020, p. 5), practically paralyzed the entire cross-border tourism business, which had generated around 20% of gross domestic product before the pandemic. Individual professional and population groups, including around 300,000 sex workers who lived on tourism, lost a large part of their income (Peer 2020, p. 5). However, in a survey in June 2020, almost 76% of Thais said they did not want foreign tourists to return to Thailand for fear that they would bring the Corona virus into the country. In contrast, only around 24% favored tourism for economic reasons (Bangkok Post, 14.06.2020).

Similar to how it remained in various neighboring countries—such as in Vietnam—the borders for foreign tourists remained closed for months in Thailand. In Thailand, even in summer 2020, the bus companies refused to transport foreigners—even if they had already settled in the country before the border closure. If you were a foreigner, you were not allowed to get on the bus—and bus employees checked Thai nationality against passports or ID cards (Bangkok Post, 13.06.2020). A journalist commented on this as follows: "Absurd, right? Don't they know that the country has been closed for months, that the airports are closed, as are the piers and border crossings since April? Under these circumstances, no foreigner or tourist is allowed to enter—where would the travel-related virus come from?" (Bangkok Post, 13.06.2020).

Tourism in Turkey collapsed by around two-thirds in 2020 after generating around US $ 35 billion in 2019. As a result of the Corona crash, around 300,000 people lost their jobs in Turkey in 2020 alone (Pabst 2021, p. 3).

According to Laura Meyer (2021, p. 2), the head of the Swiss travel company Hotelplan, bookings in November 2020 were around 20% of 2019. For the year 2021, the Hotelplan CEO expected sales of 50–60% of the pre-Corona year 2019.

In particular, countless jobs were lost in the hospitality industry during the various lockdowns in many countries in 2020 and 2021. In the USA, 21.5 million jobs or 14% of positions disappeared within 2 months in spring 2020, 6.3 million of them in the hospitality industry (Lanz 2020, p. 10). Even though a number of those who were laid off were able to return to their jobs after the restaurants reopened—the drop in income was considerable, and many employment relationships remained precarious.

Many restaurants were existentially threatened by the Corona pandemic. In Switzerland, before the Covid-19 crisis, 80% of restaurants had good to very good liquidity. By early 2021, 70% of companies were forced to submit a request for financial assistance. In 2020, 60% of all businesses had to issue termination notices (Neuhaus 2021a, p. 9). Already in the pre-Corona year 2019, 2400 restaurants disappeared from the commercial register in the Alpine Republic. Of these, around a quarter went bankrupt (Neuhaus 2021b, p. 9). By the end of 2020, 33,000 or more than 17% of the 190,000 jobs in the hospitality industry were lost in Switzerland. In March 2021, more than 60% of restaurants feared for their existence (Neuhaus 2021b, p. 9). However, in the Corona year 2020, the number of new restaurant openings in Switzerland was approximately the same as in 2019—and at the same time the number of restaurant bankruptcies and liquidations from early 2020 to April 2021 was lower than in the "normal" year 2019. Restaurants were often not closed due to bankruptcies, but because the business was abandoned or sold (Benz 2021a, p. 23). On the one hand, this shows that the government's relief measures took effect. On the other hand, apparently many entrepreneurs used the Corona forced break to implement new business ideas. According to Benz (2021a, p. 23), this also points to a "relatively normal structural change" in the Corona year.

Meher Prakash, a biophysicist who researches at the JNCAR University in Indian Bangalore and lived in the Swiss south canton of Ticino in 2021, used artificial intelligence to examine locations and measures that influenced the infection rate of Covid-

19. For the period from March to mid-September 2020, he found that restaurants with protection concepts contributed 5% to the spread of the virus, bars 3% and nightclubs 15%, with the latter proving to be ineffective. In addition, tourism contributed 18% to the spread of the virus, with the separation of travelers and commuters not always being clearly defined (Beck 2021, p. 6). If these data are correct, the tourism industry in particular—and in addition to larger events and gatherings—would have to be shut down and nightclubs closed, while the closure of restaurants and bars was rather ineffective. In fact, tourist destinations were repeatedly hotspots for the Covid-19 virus, such as the Austrian Bad Ischl in spring 2020 or the Swiss Wengen in January 2021.

In particular, the early closing times of restaurants and a series of lockdowns hit the catering industry hard in many European countries. This is because the catering industry generates around two-thirds of its sales in the evening (Schöchli 2020b, p. 11)—and also the lunch breaks fell away to a large extent in the Corona period, because many of the previous guests worked from home. The losses could not be made up for by take-away and delivery services.

There were also new trends in terms of hotel bookings. After there had been thousands of hotel cancellations due to Covid-19 in many hotels, for example, the international Radisson hotel group offered all customers free cancellation of bookings up to 24 h before arrival due to the Corona situation in December 2020 (Wildi 2020, p. 9). In general, the trend went towards shorter cancellation periods. The travel company DER Touristik offered its customers free cancellation up to 14 days before travel date until at least mid-January 2021, similar conditions were offered by providers of all-inclusive packages with charter flights. At Swiss, passengers were able to rebook reserved tickets as often and free of charge between the end of August 2020 and the end of February 2021 (Wildi 2020, p. 9). However, observers feared that, due to the thinned-out demand, the successful companies and chains could strengthen their position, which could lead to higher prices and rebooking fees in the long term.

As in Germany and Austria, the question of whether skiing should be allowed despite Corona was hotly debated in Switzerland during the second Corona wave at the end of 2020/beginning of 2021. While smaller ski resorts—such as in Central Switzerland—closed, the ski slopes in Graubünden and the Wallis remained open. In Germany and Austria, too, the ski resorts were sometimes open, sometimes closed, depending on the place and time.

In France, the ski resorts were largely closed and the ski lifts and other ski facilities incurred large losses. For example, Pascal de Thiersant, President of the ski lift company Société des Trois-Vallées, announced that his company, which had made a profit of 5 million € in 2019, wrote off 10 million € in 2020 despite state aid of approximately 50% of annual sales (Descamps 2021, p. 18). In the French ski resorts, around two thirds of the seasonal employees lost their jobs without receiving short-time work benefits (Descamps 2021, p. 18). According to the French trade union CGT, this alone affected 60,000 jobs in the Northern Alps. Almost everyone working in tourism was affected: shops, hotels, restaurants and doctor's surgeries.

The value creation in ski tourism including related services such as gastronomy, hotels and ski rental is normally around 4 billion Swiss francs per season in Switzerland. The mountain railways alone generated around 760 million Swiss francs (around 700 million euros) per winter season before Corona (Schöchli 2020a, p. 21). The number of seats in mountain railways was reduced, and as a result, the Wallis mountain railways estimated the minus in visitors compared to the previous year at −35%, in the canton of Graubünden the estimates were −17% (Benz 2021b, p. 23).

The hotel and catering industry had partly large losses to report, inter alia due to the closure of restaurants, bars and catering establishments. This was particularly the case in the Wallis and in Graubünden, where winter tourism generates around 10% of value creation (Benz 2021b, p. 23). In December 2020 and January 2021, catering in Graubünden practically came to a standstill because almost nothing was running with take away, and over the holidays the number of hotel guests was around a third below the previous year's (Benz 2021b, p. 23). In Switzerland, the Zurich region, which extends beyond the cantonal territory, recorded a 65% drop in overnight stays in 2020 compared to the previous year. In February 2021, 40% of hotels were temporarily closed (Benz and Hotz 2021, p. 24).

This was all the more serious as a large part of the turnover normally falls in January, so that in 2021 there were widespread restrictions and even closures.

In the Corona year 2020 and up to the year 2021, international tourists suddenly became a high-risk factor—and as in spring 2020, border closures were again an issue against the background of dangerous Corona mutations at the beginning of 2021. However, it was not quite without irony that almost all countries tried to exempt border crossers from possible border closures. As early as January 2021, Greece—supported by Spain and Portugal—had proposed that an obligatory vaccination passport for tourists be introduced throughout the EU in order to continue to allow cross-border tourism (Steinvorth 2021, p. 7). However, EU Council President Charles Michel said that the EU member states wanted to protect goods and commuter traffic, but not restrict non-essential travel. The resulting target conflict could only be defused by making vaccination compulsory for tourists.

But by no means all tourism sectors were equally affected by the corona-related cancellations—some even benefited from the situation. For example, Thierry Gamot, chairman of the French umbrella organization Nordic France, which represents around 200 cross-country skiing areas in France, explained that the businesses he represents achieved a sales increase of 70% in winter 2020/2021 compared to the last five years, in smaller cross-country skiing areas even 100% (Descamps 2021, p. 18). This was the result of an above-average winter, but also an effect of the fact that many downhill skiers had switched to cross-country skiing.

However, domestic tourism increased in many countries during the Covid-19 pandemic. For example, in Switzerland, in this segment, the Lucerne-Vierwaldstättersee region increased by 14%, the cantons of Wallis and Tessin by 25% and Graubünden by 36% in summer 2020 (Perren 2021, p. 19). Figure 2.13 shows the development of

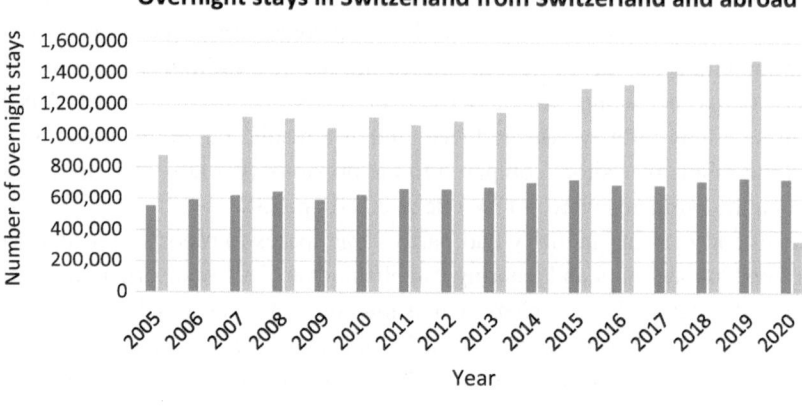

Fig. 2.13 Number of overnight stays in Switzerland by domestic and foreign guests until 2020. *Source* Nussbaumer (2021, p. 19); own representation

domestic and foreign overnight stays in Switzerland and the decline in foreign tourists in 2020.

Some countries or places pursued a "high-risk strategy" during the Corona pandemic, such as Tanzania. Up to June 2020, there were officially 509 infected and 21 dead in Tanzania, after which no more figures were published and hardly any tests were carried out. The Tanzanian President John Magufuli—who later probably died of Corona himself—declared the country free of Corona in June 2020: "We have prayed and fasted. … God has heard us," said Magufuli (quoted according to Urech 2021, p. 4). While by January 2021 already more than 100,000 people had been infected in the neighbouring state of Kenya and the second Covid-19 wave had reached its peak in Europe and the USA, tourism in Zanzibar continued as if nothing had happened: no masks, no "physical distancing", hardly any tests and no quarantines. In January 2021, full charter planes with package tourists from Eastern Europe, Russia and Ukraine arrived, on some days five planes per day from Russia and Ukraine alone. In February 2021, a music festival with 29,000 visitors was even supposed to take place on Zanzibar. However, in January 2021 the number of Corona infections in Africa was above the global average for the first time, and in the neighbouring country of Rwanda the Corona numbers exploded (Urech 2021, p. 4).

By mid-2021, the Corona situation remained unstable and unclear in many parts of the world, especially in Asia, Africa, Latin America and the Pacific, which also affected tourism. But the situation in Europe also remained uncertain and unpredictable: For example, the Corona numbers exploded on Mallorca at the beginning of the high season at the end of June 2021, at the beginning of July 2021 the more contagious Delta variant was responsible for at least 28% of the infections on the island (Schulze 2021, p. 8). As was the case earlier in the holiday areas of Portugal, the Covid-19 infections on Mallorca also increased massively—and that after Mallorca, like Portugal before, had advertised with its low Corona numbers. So in summer 2021 Germany sent all returnees from Portugal into compulsory quarantine, even vaccinated and negative test results. Therefore, tens of thousands cancelled their Portugal or Mallorca holidays in summer 2021 (Schulze 2021, p. 8).

References

Aschauer, Wolfgang 2020: Wenn die Welt zu uns spricht – Reisen im Zeitalter spätmoderner Entfremdung. In: Reif, Julian/Eisenstein, Bernd (Hrsg.): Tourismus und Gesellschaft. Kontakte – Konflikte – Konzepte. Schriften zu Tourismus und Freizeit. Band 24. Berlin: Erich Schmidt Verlag. 49 ff.

Bangkok Post 13.6.2020: „Thais-only" policy is racism, pure and simple. 14.6.2020: „Keep Foreign Tourists Out": Poll.

Beck, Renato 2021: Coronamassnahmen: Klarheit gesucht. In: WochenZeitung vom 14.1.2021. 6.

Benz, Mattias 2021a: Die Terrassenöffnungen zahlen sich aus. In: Neue Zürcher Zeitung vom 29.4.2021. 23.

Benz, Mattias 2021b: Getrübte Feiertage in den Skigebieten. In: Neue Zürcher Zeitung vom 5.1.2021. 23.

Benz, Mattias/Hotz, Stefan 2021: Schreck, lass nach! Die Schweizer Tourismusbranche will nach der Pandemie mit Nachhaltigkeit punkten. In: Neue Zürcher Zeitung vom 20.2.2021. 24.

Chen, Anze/Lu, Yunting/Ng Young C.Y. 2015: The Principles of Geotourism. Beijing/Heidelberg: Science Press/Springer.

Cohen, E. 1979: A Phenomenology of Tourism Experiences. In: Sociology. Nr. 13/2 (1979). 179 ff.

Descamps, Philippe 2021: Après Ski. Die Berge emanzipieren sich vom Pistensport. In: Le Monde Diplomatique (deutsche Ausgabe Schweiz). April 2021. 1/18/19.

Domschke, Wolfgang/Scholl, Armin 2008: Grundlagen der Betriebswirtschaftslehre. Eine Einführung aus entscheidungsorientierter Sicht. 4. Auflage. Berlin/Heidelberg: Springer.

Enzensberger, Hans Magnus 1962: Eine Theorie des Tourismus. In: Enzensberger, Hans Magnus: Einzelheiten. Frankfurt/Main: Suhrkamp. 147 ff.

Freyer, Walter 2015: Tourismus. Einführung in die Fremdenverkehrsökonomie. 11., überarbeitete und aktualisierte Auflage. Oldenbourg: De Gruyter.

Graham, Anne/Dobruszkes, Frédéric 2019: Introduction. In: Graham, Anne/Dobruszkes, Frédéric (Hrsg.): Air Transport. A Tourism Perspective. Amsterdam/Oxford: Elsevier. In: 1 ff.

Graham, Anne/Metz, David 2019: Limits to Growth. In: Graham, Anne/Dobruszkes, Frédéric (Hrsg.): Air Transport. A Tourism Perspective. Amsterdam/Oxford: Elsevier. 41 ff.

Gross, Matilde S. 2017: Gesundheitstourismus. Konstanz: UVK Verlagsgesellschaft.

Gross, Sven 2017: Handbuch Tourismus und Verkehr. Verkehrsunternehmen, Strategien und Konzepte. 2., vollständig überarbeitete und erweiterte Auflage. Konstanz/München: UVK Verlagsgesellschaft/Lucius.

Hanke, Michael 2019: Distribution Trends. In: Graham, Anne/Dobruszkes, Frédéric (Hrsg.): Air Transport. A Tourism Perspective. Amsterdam/Oxford: Elsevier. 105 ff.

Hartmann, Rainer 2020: Vorwort. In: Hartmann, Rainer (Hrsg.): Tourismus in Afrika. Chancen und Herausforderungen einer nachhaltigen Entwicklung. Oldenbourg: De Gruyter. Vf.

Helm, Roland 2009: Marketing. Strategische Analyse und marktorientierte Umsetzung. 8. Auflage. Stuttgart: Lucius & Lucius. UTB.

Heuwinkel, Kerstin 2019: Tourismussoziologie. München: UVK.

Jäggi, Christian J. 2021: Säkulare und religiöse Elemente einer ökologischen und nachhaltigen Gesellschaftsordnung. Eine Zusammenschau. Bausteine ökologischer Ordnungen. Band 5. Marburg: Metropolis.

Kaspar, Claude 1996: Die Tourismuslehre im Grundriss. 5., überarbeitete und ergänzte Auflage. St. Galler Beiträge zum Tourismus und zur Verkehrswirtschaft. Reihe Tourismus. Band 1. Bern/Stuttgart: Haupt.

Kirstges, Torsten H. 2020: Tourismus in der Kritik. Klimaschädigender Overtourism statt sauberer Industrie? München: UVK Verlag.

Lanz, Martin 2020: Millionen Amerikaner stehen vor dem Nichts. In: Neue Zürcher Zeitung vom 22.5.2020. 10.

MacCannell, Dean 1973: Staged Authenticity. Arrangements of Social Space in Tourist Settings. In: American Journal of Sociology. 79/3 (1973). 589 ff.

MacCannell, Dean 1999: The Tourist. A New Theory of the Leisure Class. Berkley: University of California Press.

Meyer, Laura 2021: „Die Buchungen nehmen zu". Gespräch mit Laura Meyer, Chefin von Hotelplan. Von Benjamin Weinmann. In: Luzerner Zeitung vom 13.4.2021. 2f.

Mihai, Silviu 2021: Menschenleere Paradiese. In: WochenZeitung vom 21.1.2021. 10.

Müller, Hansruedi 2011: Tourismuspolitik. Wege zu einer nachhaltigen Entwicklung. Glarus/Chur: Rüegger.

Mundt, Jörn W. 2013: Tourismus. 4., überarbeitete und ergänzte Auflage. München: Oldenbourg Verlag.

Neuhaus, Christina 2021a: Jedes zweite Restaurant sieht sich bedroht. In: Neue Zürcher Zeitung vom 12.1.2021. 9.

Neuhaus, Christina 2021b: 33.000 Gastro-Jobs weg. In: Neue Zürcher Zeitung vom 27.3.2021. 9.

Newsome, David/Moore, Susan A./Dowling, Ross K. 2013: Natural Area Tourism. Ecology, Impacts and Management. 2nd Edition. Bristol/Buffalo/Toronto: Channel View Publications.

Nussbaumer, Lukas 2021: S. 19: Schweizer Gäste polieren die miese Bilanz etwas auf. In: Luzerner Zeitung vom 7.4.2021. S. 19.

Pabst, Volker 2021: Der Tourismus hat oberste Priorität in der Türkei. In: Neue Zürcher Zeitung vom 12.5.2021. 3.

Page, Stephen J./Connell, Joanne 2014: Tourism. A Modern Synthesis. Fourth Edition. Hampshire/UK: Cengage Learning.

Peer, Mathias 2020: Das Coronavirus legt in Thailand das Sexgewerbe lahm. In: Neue Zürcher Zeitung vom 20.11.2020. 5.

Perren, Marcel 2021: „Es wird weniger Gruppen geben". Interview mit dem Luzerner Tourismusdirektor Marcel Perren von Stefan Dähler. In: Luzerner Zeitung vom 12.1.2021. 19.

Plüss, Christine 2019: „Reisen ist ein kostbares Gut". Gespräch mit Christine Plüss von Susanna Müller. In: Neue Zürcher Zeitung vom 15.3.2019. 55.

Putz, Ulrike 2020: Balkonien statt Binnentourismus in Japan. In: Neue Zürcher Zeitung vom 21.12.2020. 3.

Rate, Shirley/Moutinho, Luiz/Ballantyne, Ronnie 2018: The New Business. Environment and Trends in Tourism. In: Moutinho, Luiz/Vargas-Sánchez, Alonso (Hrsg.): Strategic Management in Tourism. 3rd Edition. Oxfordshire/Boston: Cabi. 1 ff.

Rosdorff, Julia 2020: Der Einfluss des Afrika-Images auf die Reiseentscheidung der Deutschen: Probleme und Herausforderungen für das Tourismusmarketing afrikanischer Staaten. In: Hartmann, Rainer (Hrsg.): Tourismus in Afrika. Chancen und Herausforderungen einer nachhaltigen Entwicklung. Berlin/Boston: Walter de Gruyter. 50 ff.

Saarinen, Jarkko 2014: Transforming Destinations: A Discursive Approach to Tourist Destinations and Development. In: Viken, Arvid/Granås, Brynhild (Hrsg.): Tourism Destination Development. Turns and Tactics. Farnham: Ashgate. 47 ff.

Schäfer, Robert 2015: Tourismus und Authentizität. Zur gesellschaftlichen Organisation von Außeralltäglichkeit. Kulturen der Gesellschaft. Band 14. Bielefeld: Transcript.

Schiele, Kertin 2017: Tourismus und Identität. Vietnam-Reisen als Identitätsarbeit von in Deutschland lebenden Việt Kiều. Berlin: regiospectra.

Schnorbus, Linda/Wachowiak, Helmut 2020: Urlaubsglück der Deutschen: Was macht die Deutschen im Urlaub glücklich? In: Reif, Julian/Eisenstein, Bernd (Hrsg.): Tourismus und Gesellschaft. Kontakte – Konflikte – Konzepte. Schriften zu Tourismus und Freizeit. Band 24. Berlin: Erich Schmidt Verlag. 25 ff.

Schöchli, Hansueli 2020a: Mehr Nothilfe für Betriebe. In: Neue Zürcher Zeitung vom 10.12.2020. 21.

Schöchli, Hansueli 2020b: 1,5 Milliarden Franken zusätzlich für Nothilfen. In: Neue Zürcher Zeitung vom 12.12.2020. 11.

Schulze, Ralph 2021: Delta breitet sich aus – „Mallorca steht am Rande des Abgrunds". In: Luzerne Zeitung vom 3.7.2021. 8.

Scott, Daniel/Halland, C. Michael/Gössling, Stefan 2012: Tourism and Climate Change. Impacts, Adaptation and Mitigation. London/New York: Routledge.

Statista 2021a: Market size of the global hotel industry from 2014 to 2018. https://www.statista.com/statistics/247264/total-revenue-of-the-global-hotel-industry/ (Zugriff 8.1.2021).

Statista 2021b: Weltweites Tourismusaufkommen nach Reiseankünften bis 2019. Veröffentlicht von Lena Gaefe. 5.1.2021. https://de.statista.com/statistik/daten/studie/37123/umfrage/weltweites-tourismusaufkommen-nach-reiseankuenften-seit-1950/ (Zugriff 7.1.2021).

Statista 2021c: Weltweite Tourismuseinnahmen von 2000 bis 2018. https://de.statista.com/statistik/daten/studie/187764/umfrage/weltweite-einnahmen-im-tourismus-seit-2000/ (Zugriff 7.1.2021).

Steinbach, Josef 2003: Tourismus. Einführung in das räumlich-zeitliche System. München/Wien: Oldenbourg.

Steinvorth, Daniel 2021: Das Gespenst des Grenzchaos geht um. In: Neue Zürcher Zeitung vom 23.1.2021. 7.

Tschopp, Martin/Beige, Sigrun/Axhausen, Kay W. 2011: Forschungsbericht NFP 48: Verkehrssystem, Touristenverhalten und Raumstruktur in alpinen Landschaften. Zürich: vdf Hochschulverlag.

UNWTO 2018: World Tourism Organization: UNWTO Tourism Highlights 2018 Edition. www.e-unwto.org/doi/pdf/10.18111/9789284419876 (Zugriff 8.3.2021).

Urech, Fabian 2021: Das Märchen von Sansibar als Paradies ohne Corona. In Neue Zürcher Zeitung vom 1.2.2021. 4.

Urry, John/Larsen, Jonas 2011: The Tourist Gaze 3.0. Thousand Oaks: Sage.

Vester, Heinz-Günter 1999: Tourismustheorie. Soziologischer Wegweiser zum Verständnis touristischer Phänomene. München/Wien: Profil.

Wildi, Robert 2020: Kampf um die Stornogebühren. In: Luzerner Zeitung vom 28.12.2020. 9.

Economic Importance of Tourism

3

Freyer (2015, p. XV) has pointed out that since 1990, research and teaching in tourism have undergone a wide development. On the one hand, the basic principles of tourism have been further developed and substantiated scientifically, and tourism is represented as its own research and teaching area at many universities. On the other hand, the tourist markets have grown rapidly, the market volume has reached an astronomical level, especially in terms of holiday travel, air travel and package tourism, making a collapse almost inevitable, as with the Covid-19 pandemic. Although tourist demand in many industrial countries had been stagnating at a high level before Covid-19, and digital media have been increasingly changing travel since the turn of the millennium, new demand groups have emerged, especially in Asia, Eastern Europe and partly in Latin America. At the same time, the tourist offer has become more differentiated and also more competitive. Many tourism companies have reacted and are reacting more professionally and innovatively to changes in the markets (Freyer 2015, p. XV).

Tourism as a system can be represented diachronically as well as synchronously (Fig. 3.1).

Tourist offers can be broken down according to various criteria, including destination type (Table 3.1).

Mayer and Vogt (2016, p. 101) have listed the most important determinants of tourist spending in the Alpine region with a view to nature-based tourism (Table 3.2).

Revenue per traveler is also determined by travel discounts. For example, an employee of a shipping company in Switzerland once explained to me that, due to package deals and special subscriptions, Asian tourists generate an average of only one tenth of the revenue of a local tourist, for whom there are no corresponding discounts. This could also be seen as a form of discrimination against the local population …

The price elasticity of demand for tourism has decreased significantly over the years: In the 1960s, the demand for tourism was similar to the demand for luxury goods.

C. J. Jäggi, *Tourism Before, During and After Corona*,
https://doi.org/10.1007/978-3-658-39182-9_3

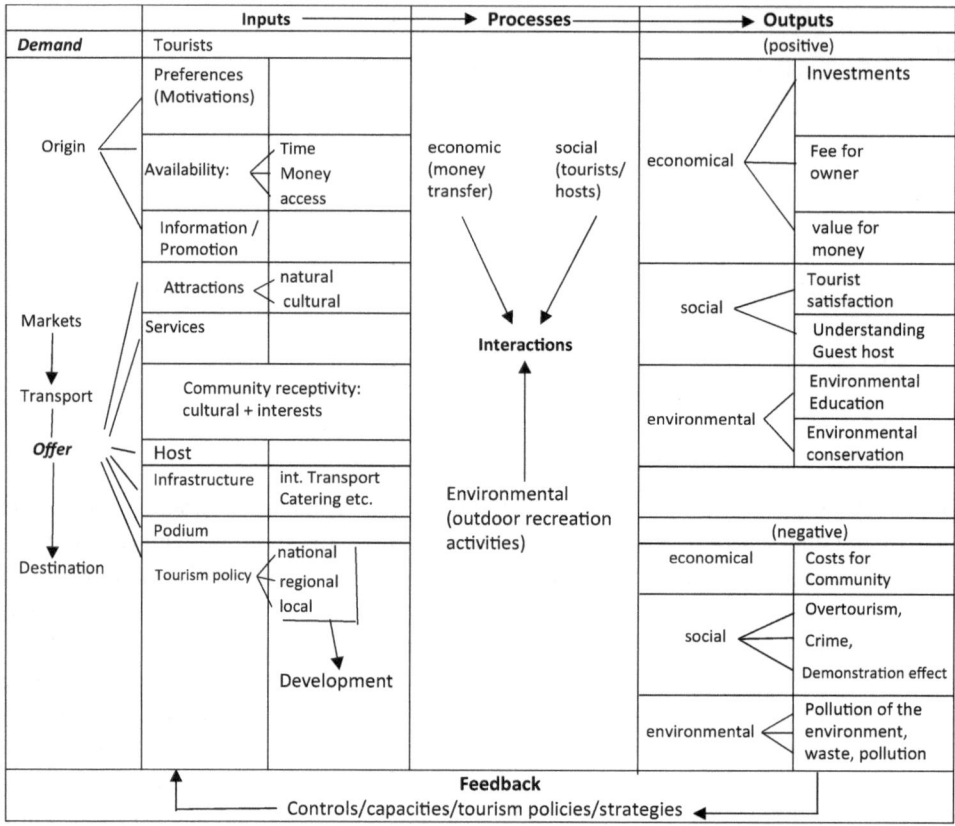

Fig. 3.1 Tourism system. *Source* Newsome et al. (2013, p. 11), slightly modified and translated from English by the author

Table 3.1 Destination types. (Aus Letzner 2014, S. 7)

Destination types	Non-tourist uses	Tourist infrastructure	Tourist attractions
a) **Natural landscape**	None	None to very low	Traditional
b) **Cultural landscape**	Medium	Medium	Predominantly traditional
c) **Cityscape**	High	Medium to high	Everything, from traditional to produced
d) **Ski or hot springs destination**	Medium	Medium to high	Traditional (agricultural forms and climate); partly also produced
e) **Artificial (leisure) spaces**	Medium to high	Very high	Produced
For comparison: an industrial and commercial region	Very high	None (possibly hotels and exhibition halls)	None

Table 3.2 Determinants for tourist spending behavior. (Mod. according to Mayer and Vogt 2016, p. 101; supplemented by the author; own representation)

Determinants of the/the		
Traveler	Travel	Destination
Age	Number of traveling companions	Distance to the source area
Gender	Travel duration	Perceived price level
Income	Overnight versus day guests	Actual price level
Family relationships	Type of accommodation	Assessment of holiday/trip
Education	Travel organization	Characteristics of the destination (location, infrastructure, offer …)
Profession	Means of transport (on arrival and on site)	Season
Country of origin	Number of visits	Climate
Residence category	Travel motives	Weather
…	Activities	…
	…	

Table 3.3 Development of elasticity in demand for overnight stays in international travel between 1960 and 1985 in relation to gross domestic product and real income. (From Kaspar 1991, p. 119)

Origin	1960	1970	1980	1985
Austria	1.24	1.14	1.21	1.17
France	1.10	1.06	1.26	1.21
BRD	2.45	1.50	1.32	1.26
Switzerland	1.33	1.19	1.10	1.09
United Kingdom	3.46	2.05	1.96	1.79
Belgium	1.47	1.26	1.20	1.16
Netherlands	1.89	1.38	1.26	1.20
Sweden	2.22	1.72	1.56	1.43
USA	2.46	1.60	1.71	1.55

This changed at the latest in the 1990s, when the income elasticity of tourism demand assumed values practically everywhere below 1. Since then, changes in income have had only very little effect on tourism demand. Higher income elasticity exists for foreign offers. Table 3.3 shows this development between 1960 and 1985.

It is undisputed that tourism has many positive economic and social impacts and can also lead to an improvement in the local environmental situation. These include, among other things, the creation of location-related jobs, the provision of jobs, especially for those with little qualifications, direct and indirect regional economic effects, development impulses for less developed areas, diversification of the economy, for example in purely agricultural areas, positive effects on the preservation of local cultural resources, the construction of infrastructure facilities such as sewage treatment plants, etc. (Strasdas 2017, pp. 19 ff.). However, all these advantages can quickly turn into their opposite if they are used excessively or one-sidedly for commercial purposes: Dumping wages can be paid at tourism-specific jobs, the additional wage of those with little qualifications can be absorbed by price increases in the food sector and tourism can lead to local traffic congestion or turn into overtourism. Also, tourism can marginalize or destroy other economic sectors by too strong a focus on imports. Too one-sided tourist orientation can lead to de-diversification of the local economy, sustainable subsistence agriculture disappears, the local culture becomes tourist folklore and the construction of infrastructure projects is limited to the immediate vicinity of tourist facilities or even has negative effects, such as excessive private traffic as a result of better roads.

Wattanakuljarus and Coxhead (2013, pp. 182 ff.) examined the effects of tourism on poverty in Thailand. They concluded that while all income classes benefit from tourism growth, tourism benefits households with high incomes and non-agricultural households the most (Wattanakuljarus and Coxhead 2013, p. 205). As long as agricultural households and other labor-intensive sectors are not involved in tourism, it is difficult to speak of a poverty-reducing effect of cross-border tourism. At the same time, tourism leads to increased income inequality and increasingly divergent returns on investment (ROI) of individual economic sectors. In addition, tourism has contributed to enormous increases in property prices, especially at tourist hotspots.

Bruno S. Frey (2020, p. 17), an economist of culture, has also pointed out that today—after the positive effects of tourism on the economy were long undisputed—a more critical view of tourism and especially mass tourism is gaining ground.

3.1   Global

In 2018, tourism was the second-fastest growing economic sector in the world. Its share of world GDP was 10.4% (Muntschik 2019, p. 10).

Freyer (2011, p. 51) has summarized the business, micro- and macroeconomic aspects of tourism in a schema (Fig. 3.2).

From the perspective of markets, tourism can be divided into various submarkets (see Fig. 3.3).

As a result of globalization, the individual tourism sectors have developed rapidly in the last 20 years. Table 3.4 shows the world's largest hotel chains in 2007.

In the past decade, the number of cruise passengers has continued to rise, until the collapse in the Corona year 2020 (Fig. 3.4).

In 2019, of the approximately 30 million people who took a cruise, around 10 million were from the USA and just under 3 million from Germany (Kirstges 2020, p. 32). From

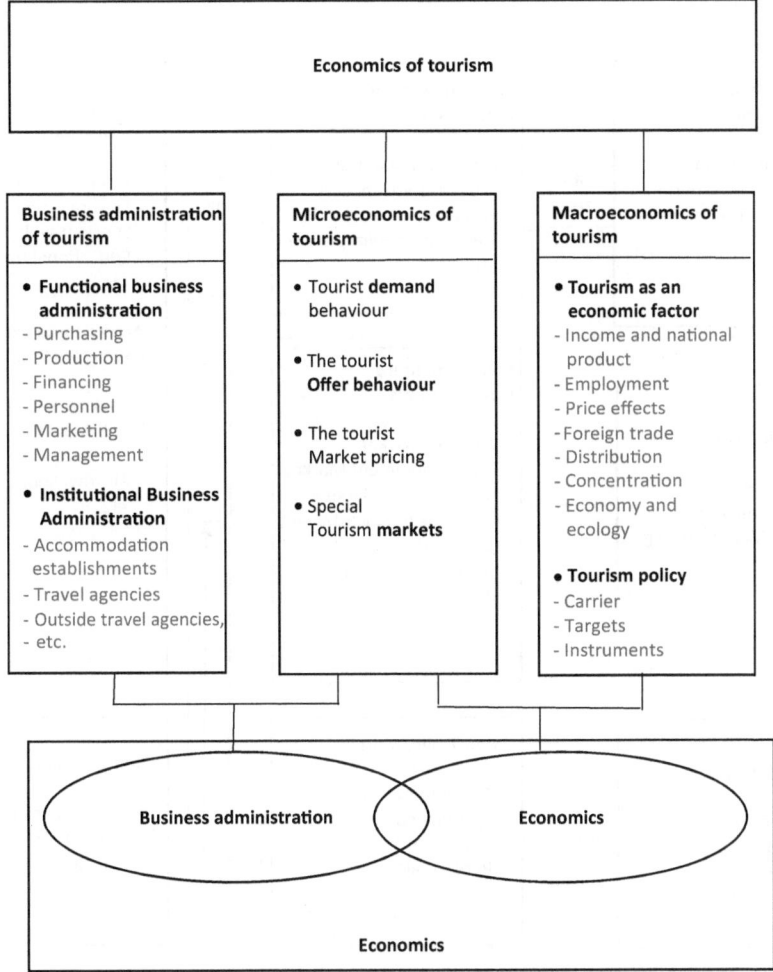

Fig. 3.2 The three pillars of an economy of tourism. (Mod. after Freyer 2011, p. 51; slightly modified and simplified representation by the author)

2019 to 2020, the number of cruise passengers from Germany fell from 3.124 million to 1.405 million, after steadily rising since 2004 (Statista 2021).

According to Schulz (2014, p. 69), the Mediterranean was the most popular destination for German ocean cruise passengers (37%), followed by the Nordic countries (20%), the USA and the Caribbean (15%), Western Europe (12%), the Baltic Sea (10%) and other overseas areas (6%).

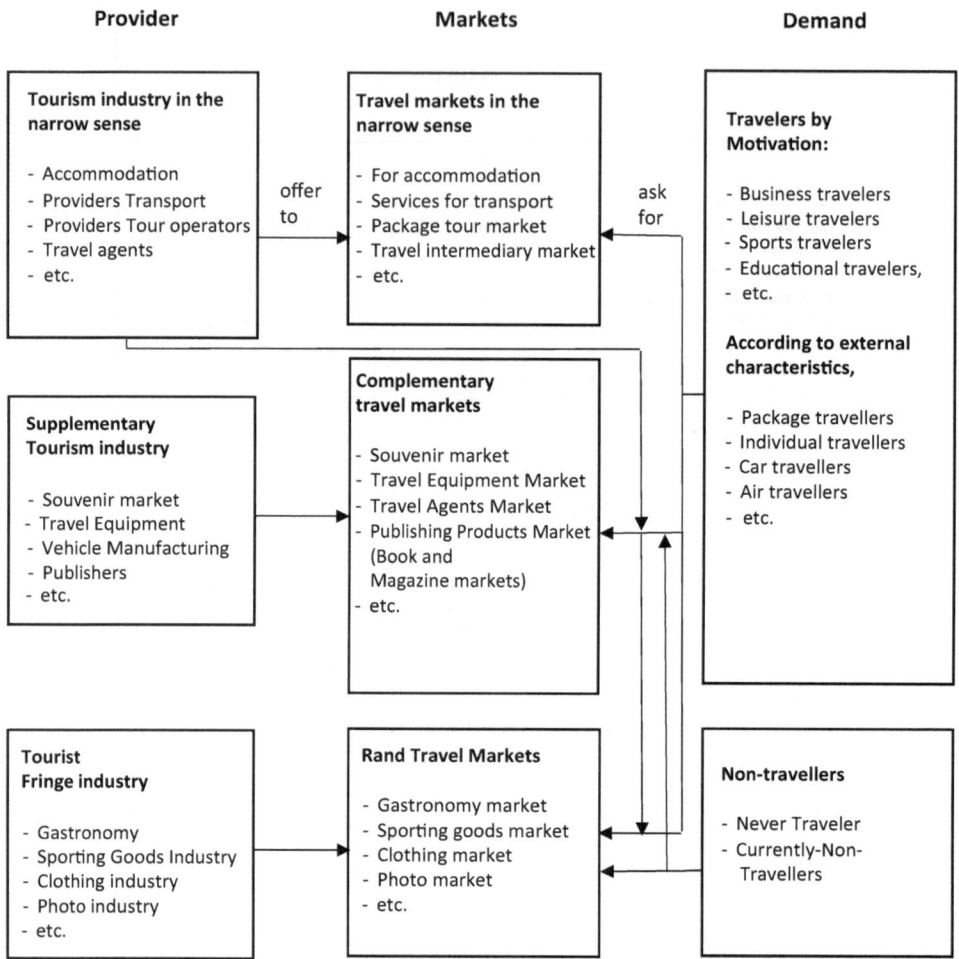

Fig. 3.3 Suppliers, markets and consumers in tourism. (Mod. after Freyer 2011, p. 307, slightly simplified and modified by the author)

3.2 National

In many countries, tourism is one of the main pillars of the economy.

As Fig. 3.5 shows, in 2019 Spain, France, Italy and Germany earned the most from tourism in Europe before the Corona year.

In terms of gross domestic product, the importance of tourism in Europe was and is greatest in Croatia: in 2018, the share of tourism in GDP in Croatia was 11%, in Greece 8.3%, in Italy 5.5%, in Spain 5.4%, in France 3.7% and in Switzerland 2.5% (Müller et al. 2021, p. 25). Indirect effects of tourism on the economy are not taken into account.

Table 3.4 The world's largest hotel chains measured by the number of rooms. (From Clancy 2011, p. 82;)

Rank	Company	Country	Number of rooms	Number of properties
1.	IHG (InterContinental)	United Kingdom	585,094	3949
2.	Wyndham Hotel Group	USA	550,576	6544
3.	Marriott International	USA	537,249	2999
4.	Hilton Hotels Corp	USA	502,116	3000
5.	Accor	France	461,698	3871
6.	Choice Hotels International	USA	452,027	5570
7.	Best Western International	USA	308,636	4035
8.	Starwood	USA	274,535	897
9.	Carlson Hotels Worldwide	USA	146,600	969
10.	Global Hyatt Corp	USA	135,001	721

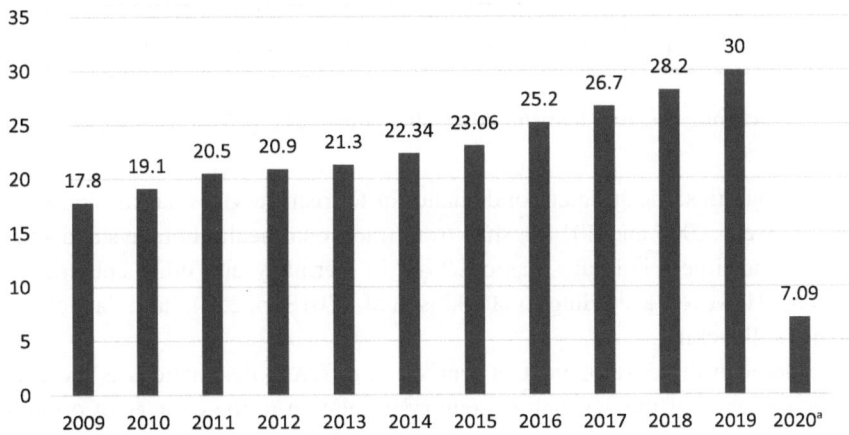

Number of cruise passengers worldwide per year in millions

Fig. 3.4 Cruise passengers per year. (From Jordan 2021, p. 15; own representation. [a] For the year 2020, the estimate from Wikipedia was used due to lack of accurate numbers[1])

---

[1] https://de.wikipedia.org/wiki/Kreuzfahrt (accessed 10.07.2021). For comparison: Statista had estimated the number of cruise tourists for the year 2020 before the Corona crisis at 32 million travelers (https://de.statista.com/statistik/daten/studie/285194/umfrage/passagiere-auf-dem-welt-weiten-kreuzfahrtmarkt/ accessed 10.07.2021).

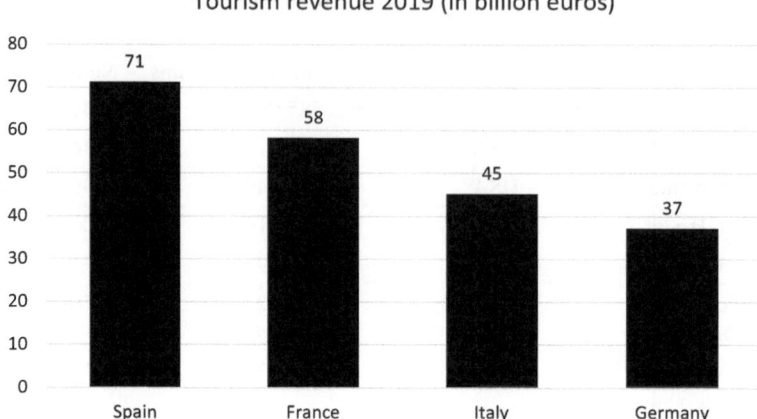

Fig. 3.5 Tourism revenue in Spain, France, Italy and Germany. (From Schmutz et al. 2020, p. 18; own representation)

Fig. 3.6 Product components in tourism. (Mod. after Gross 2017, p. 77; own representation)

In many countries, the structure of demand for tourism services has also changed. For example, between 2003 and 2014, a shift from traditional health holidays and spa offers to wellness and fitness holidays was observed in Germany and other countries (Gross 2017, p. 65). However, according to Hopkins et al. (2018, p. 252), hard data on medical tourism are still lacking.

A key factor in the development of tourism at different destinations is the exchange rate of the currency of the destination country in relation to the country of origin of the tourists. For example, since 2000, the number of overnight stays by foreign tourists in Switzerland has developed in parallel with the exchange rate of the Swiss franc—and vice versa, the booking of foreign holiday destinations by Swiss tourists (Zenhäusern and Kadelbach 2018, p. 10).

3.3   Local

In tourism, various product components can be listed with respect to location (Fig. 3.6).

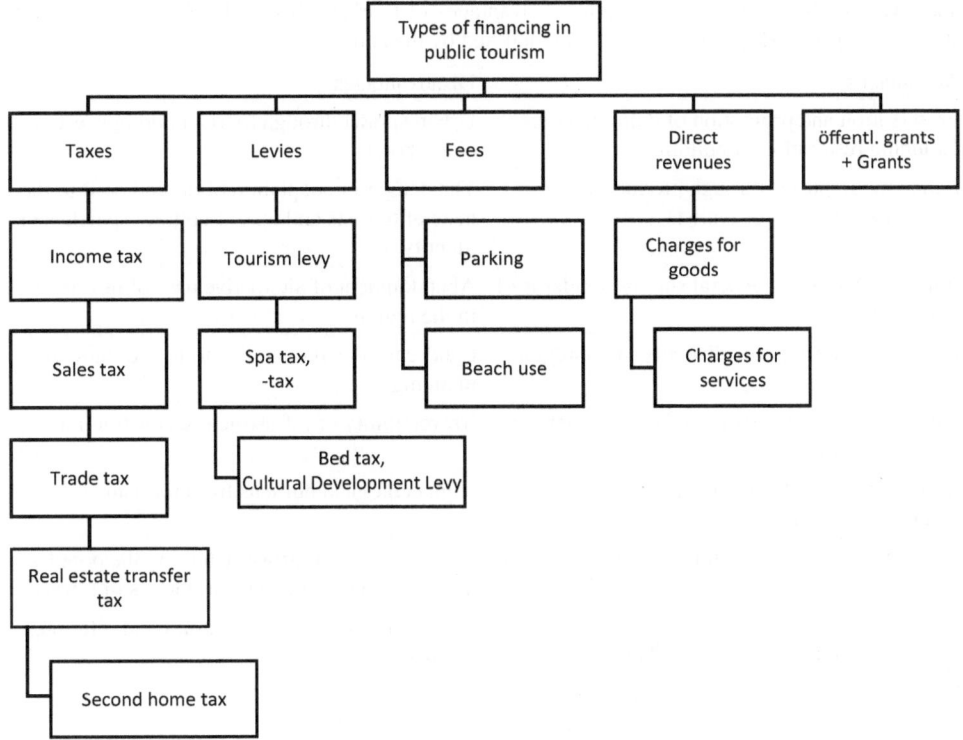

Fig. 3.7 Taxes, fees and charges to national and local authorities. (From Gross 2017, p. 146; own representation)

At the same time, tourism not only generates revenue for the mostly private tourism providers, but also for local authorities. Gross has schematically represented the possible or actual taxes, fees and charges in Germany (Fig. 3.7).

3.4   Positive and Negative Impacts of Tourism

Tietz (1980, p. 189) already listed eight advantages and as many disadvantages of tourism 40 years ago (Table 3.5).

According to Kirstges (2020, pp. 15 ff.), tourism has the following positive effects, advantages and opportunities: In addition to economic benefits such as value creation, job creation and prosperity in source and destination countries, tourism promotes understanding between peoples, promotes social development in destination areas, helps to build infrastructure in destination areas, supports environmental and species protection through the preservation of "intact nature", promotes awareness *("awareness")* of risks

Table 3.5 Eight advantages and eight disadvantages of tourism. (Mod. after Tietz 1980, p. 189; cit. after Kaspar 1991, p. 114; adapted and modified by the author)

Advantages	Disadvantages
Preservation and promotion of cultural assets, including restoration of buildings	Urban sprawl through tourist buildings, especially hotels
Economic impulses through production of goods and services for tourists	The danger of adaptation of art and craft to the taste of tourists and loss of cultural specifics or identity
Improvement of the general supply standards of the population	Abandonment of alternative uses of resources in the region
Expansion of economic diversity in the region	Landscape destruction due to lack of burden planning
Part-time offers and additional employment in agricultural areas	Irreversibility of infrastructure construction
Improved regional and local infrastructure thanks to tourism	High ecological burdens from tourism
Use of tourism infrastructure by the population	No use of tourist infrastructure by the population due to lack of means or ghettos of tourists
Higher awareness of the region, cultural encounters and new patterns of behavior at the local level	Disturbance of the local cultural and religious balance

and dangers to human, animal and plant life and makes people happy through travel. Each of these points could be challenged—but it is only undisputed that tourism generates a significant contribution to the gross domestic product of individual countries and creates jobs—but often only precarious jobs, seasonal employment opportunities and not always full-time positions (Kirstges 2020, p. 27). All other points could also be seen as euphemisms or as "green washing" from a critical perspective: Because tourism promotes understanding between peoples as well as resentment. The social development in many destination areas often consists rather in a separation of the tourists from the local society or even in a growing social polarization. The infrastructure is often overloaded rather than expanded, nature reserves are often no more than suppliers of photo subjects and are often exhausted and overloaded by the tourist streams, millions of tons of harmful emissions are emitted during the journey to and from the destination—especially by airplane—and the happiness gained during the journey is often only a temporary escape from everyday problems.

Kirstges (2020, pp. 27 ff.) has also dealt extensively with the negative aspects of tourism.

All this basically means that tourism has numerous positive as well as many negative aspects, which is why positive aspects should be promoted and negative ones minimized as far as possible by means of a balanced tourism policy.

3.5   Economic Problems in Tourism

Like all economic sectors, tourism is also confronted with specific difficulties and problems.

In many places, the offerings and infrastructure are still geared towards peak times, many hoteliers and restaurateurs only earn money for a few months of the year—the rest of the time they live on reserves. If the high season collapses, it can sometimes take years for tourism providers to recover. When politicians or interest representatives demand the promotion of tourism, they often only have the capacities oriented towards the peak times in mind. Imwinkelried (2020, p. 10) states: "In the mountains, the season has become shorter and shorter in recent years. Many hotels are only open for four months in winter and just as long in summer". Accordingly, it is difficult to find qualified staff. And if, in addition, exactly the densest travel and tourism time is lost due to external influences, as was the case in winter 2020/2021 or partly in summer 2020 due to the Covid-19 pandemic, this quickly leads to existential problems in the industry.

That is why innovative companies—such as Andermatt Swiss Alps (ASA), which belongs to the Orascom Group—have developed new strategies: For example, the Swiss Andermatt is to be converted into an integrated year-round destination that offers first and second homes for Swiss and foreign citizens—not only a place for sports and leisure, but also for living, working and networking with the locals (Imwinkelried 2020, p. 10). However, it is only possible to sustainably increase the value added generated by the tourism service providers if the guests stay longer. However, there are two problems with this: First, the local infrastructure must not be too large and the burden on the natural environment must not be too great, and second, the propagation of second homes is also not unproblematic, because this results in an disproportionate ecological footprint[2].

3.6   Tourism generates Jobs

It is undisputed that tourism generates many urgently needed jobs in many countries and regions. Many of these jobs are open to women, young people, poorly qualified people and part-time employees, unlike other economic sectors. According to Elshaer and Marzouk (2020, p. 67), many self-employed people work in the tourism sector, as well as many small businesses. However, many tourist jobs are low-wage, but often only require low qualifications. Many jobs in tourism and hospitality are seasonal, often also involving people from the secondary labour market[3] (Elshaer and Marzouk 2020, p. 67).

---

[2] For the ecological footprint, see also footnote 2 (Chap. 4).

[3] The secondary or second labour market includes all non-employment or reduced-wage, comparable or similar activities, which are practically always organised by an institutional provider and are carried out outside one's own four walls in most cases. The state-subsidised second labour market is particularly well-known in the general public, but in the strict sense it is not a market, but includes subsidised employment opportunities and measures (based on https://de.wikipedia.org/wiki/Zweiter_Arbeitsmarkt Accessed 09.01.2021).

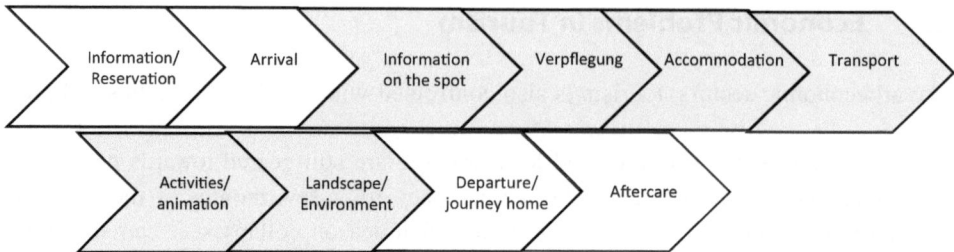

Fig. 3.8 Service chain in tourism. (Mod. after Dorsch 2016, p. 35; Müller 2011, p. 65; FIF 1995; slightly edited by the author, own representation)

Figure 3.8 shows the tourist service chain.

In the service sector—and thus also in tourism—wage costs account for more than 70% of total costs. Even more than in other industries, the quality of work—that is, human capital—is decisive for business success (Cook et al. 2018, p. 91).

On cruise ships, the ratio between crew members and guests—the so-called *"passenger-crew-ratio"*—is between 1:0.25 and 1:1, that is, 1 to 4 passengers per employee (Kirstges 2020, p. 34). In addition, around 70% of the crew usually come from Southeast Asia, Central America and the Caribbean, 30% or more are Filipinos (Kirstges 2020, p. 34).

3.7   New Ways of Working

In the light of new developments in the world of work, the question arises as to what effect these changes will have on tourism. For example, since the end of the 1990s, the number of self-employed people in Germany has increased by more than 40% from around 3 to 4.3 million (Merkel 2018, p. 33). The "new self-employed" differ from traditional entrepreneurs mainly in their above-average educational qualifications, a higher proportion of women and migrants, and their work as "single self-employed" (Merkel 2018, p. 33). They often suffer from deficits in social rights, social security and in relation to their professional interest representation. As a result of the structural change in work society due to the erosion of normal employment relationships, increasing individualization and the pluralization of contract forms, the new employment relationships appear to be more flexible and better adapted to individual needs, but they are usually also more precarious, more risky in terms of future developments and sometimes also worse paid.

In the tourism and accommodation sector too, new forms of work and services have been established, for example via digital platforms such as Airbnb, the expansion of accommodation options such as bed and breakfast, couchsurfing, etc. In addition, there is an increasing number of so-called "digital nomads", that is, people who set up their workplace somewhere in the world and, thanks to the internet, notebook and smartphone, can and want to work anywhere.

3.8   Tourism as a Cause of Dual Markets and Inflation

With regard to the Inca Trail in Peru, Maxwell (2012, p. 115) found that tourism does not necessarily lead to the locals "losing their culture"—but tourists are often the cause of a dual price development. For example, village residents in Peru tried to protect their village markets from the uncertainties of the tourism industry by charging significantly higher prices from tourists than from locals. The practice of taxi drivers in many countries, for example in Turkey or Portugal, of charging greatly inflated prices from foreign tourists or no longer accepting shorter taxi rides, but, for example, only driving tourists to the Algarve in Lisbon, is also well known. Other examples are carpet dealers who offer oriental carpets to western tourists at greatly inflated prices. Craftsmen offer their work for a significantly higher price in many places than for local customers (Maxwell 2012, p. 115).

Kirstges (2020, p. 28) has pointed out that tourism-induced inflation can become a problem. In many tourist destinations, the price level increased with the growing number of tourists. Many goods became unaffordable for locals who did not benefit from the tourism boom. In many tourist destinations, the prices of land and property rose rapidly as more and more people from outside the area bought property there as a second home, such as in Barcelona. On the Ile de Ré off the French Atlantic coast, the price of land rose sharply in the 1990s and 2000s as many outsiders bought property there as a second home. As a result, the value of the land of the locals also rose, many of them became liable for wealth tax—Impôt de Solidarité sur la Fortune—but were unable to pay this tax due to insufficient income (Kirstges 2020, p. 28). Similar developments took place in other parts of France, in Great Britain and in many other places.

3.9   Economic Perspectives for Tourism After Corona

If it is true that the desire to travel is unbroken—as for example the CEO of Migros[4] Fabrice Zumbrunnen said (Zumbrunnen 2021, p. 25)—and people only hesitate to book holidays in spring 2021 due to the unclear travel conditions and -possibilities, it is to be expected that travel activity will increase again after the Corona pandemic. However—as Zumbrunnen (2021, p. 25) said—the travel industry expected that business travel in the short term would decrease or remain low.

---

[4] The Migros Group is the second largest retailer in Switzerland after Coop.

In the middle of the Covid-19 crisis, when asked what the Corona pandemic would do to the future of tourism, André Lüthi (2020, p. 49), CEO of the Globetrotter Group, to which 14 travel companies in Switzerland belong, said: "First of all, during and after the pandemic, 'experience and discovery tours, but also political, study, wellness, bike and language tours' will be in demand again. However, after an initial phase with price discounts, higher prices are to be expected. Niche market providers are likely to have the best chance of surviving the crisis. Lüthi also expected a wave of bankruptcies among lodges, car rental companies, airlines and hotels. Lüthi (2020, p. 49) also said that people might travel more consciously in the future and make fewer but longer trips instead of many short city trips.

Many tourist destinations suffered massively from the Corona pandemic. In 2019, 67 million tourists visited the city of New York, in 2020 only 23 million (Petrin 2021, p. 6). In 2019, tourists in New York spent $47.4 billion, making tourism one of the most lucrative branches of the New York economy. A recovery of tourism was not expected in New York until 2025 (Petrin 2021, p. 6).

In Switzerland, a trend that had already been apparent before was reinforced during the Corona period: While the number of overnight stays in urban regions increased significantly from 2007 to 2017, they decreased just as clearly in mountain regions—for example in Graubünden, in the Valais, in the Bernese Oberland and in Ticino. The whole situation is even more problematic for mountain regions because the value added by tourism in mountain regions makes up a significantly higher proportion of gross value added than in larger cities[5] (Zenhäusern and Kadelbach 2018, pp. 10–11). Figure 3.9 compares mountain regions with large cities and other areas.

The corona crisis has made and exacerbated existing structural problems in the tourism sector visible. For example, in summer 2020, occupancy rates in city hotels in Switzerland were often only 10–20%. Although domestic tourism made up for some of the losses of foreign tourists in mountain regions, many hotels had serious problems. For example, it has been estimated in Switzerland for some time that around one third of hotels in mountain regions are not viable—and there are similar estimates for Austria (Benz 2020, p. 15). While hotels in cities can be easily converted and rented out as living space, the situation in mountain regions is more difficult (Benz 2020, p. 15).

In the cruise sector, Pythagoras Nagos (2020), former managing director of MSC Cruises in Switzerland, expected in the middle of the first corona wave in May 2020 that the occupancy rate on cruise ships would drop to 40–50% due to the demand for "physical distancing", and he even discussed a temporary travel ban for people over the age of 70.

---

[5] The number of people directly employed in tourism was 16.1% in mountain regions of Switzerland in 2015, and those indirectly employed in tourism 10.7% (measured in full-time equivalents), compared with 4.5% and 2.7% in large cities; see Fig. 3.9 . The share of tourism in gross value added was 20.8% in mountain regions in 2015 (of which 13.1% directly and 7.7% indirectly), 4.7% in large cities (of which 3.1% directly and 1.6% indirectly) and 3.2% in other areas (1.9% directly and 1.3% indirectly) (Zenhäusern and Kadelbach 2018, p. 11).

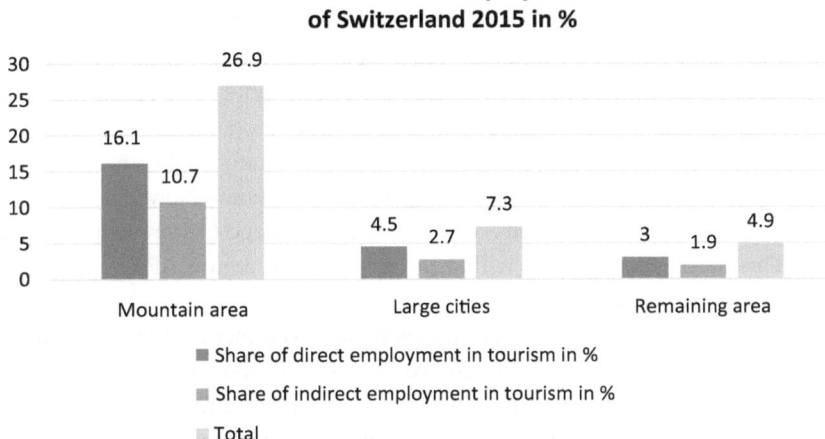

Fig. 3.9 Tourism employment (VZA) in Switzerland in mountain regions and larger cities in 2015. (From Zenhäusern and Kadelbach 2018, p. 11; own representation)

The aviation sector and, consequently, the aviation industry were particularly affected by the 2020 coronavirus crisis. In the first half of 2020, production in the global aviation industry collapsed by 50% overnight, and in summer 2020 Airbus announced the cancellation of 15,000 jobs (Descamps 2020, p. 10). In France, where there were more than 300,000 direct jobs in the aviation industry in 2019 and this sector was by far the leader in terms of turnover, foreign trade volume and research, the collapse was enormous. In addition, the aviation sector has hardly any technical alternatives to harmful greenhouse gas emissions, in contrast to other industrial sectors (Descamps 2020, p. 10).

Airports were also severely affected by the coronavirus pandemic. However, hardly anyone had anticipated how long the difficult situation would last in many places. While Zurich Airport had achieved a turnover of CHF 1210 million in 2019—that is, before the coronavirus—this figure had fallen to CHF 590.7 million by 2020—and for 2021, forecasts in March of that year predicted 838.2 million (Müller 2021, p. 1)—which was probably too optimistic. Because there were already signs of a third coronavirus wave and in March 2021, the WHO warned of a trend reversal in Covid-19 infections, as the number of infections had risen for the first time in 6 weeks after 6 weeks of declining infection rates.

In the opinion of Christian H. Kaelin, Chairman of Henley & Partners and the inventor of the Passport Index concept, the travel options for holders of different passports vary depending on the situation and general conditions. This was also the case during the coronavirus pandemic in 2020 and 2021. "In the last 7 years, the US passport has fallen from number one to number 7, a position it currently [June 2021] shares with the UK. Due to pandemic-related travel restrictions, travelers from both countries are currently confronted with significant restrictions in over 105 countries, with US passport holders

able to travel to fewer than 75 countries, while UK passport holders can currently travel
to fewer than 70 countries" (Tageskarte 2021; note by the author).

Aside from temporary restrictions, Japan held the top position on the Passport
Index—holders of a Japanese passport could visa-free travel to 191 countries as of early
2021. In second place was Singapore with 190 reachable countries, followed by South
Korea and Germany in third place with 189 reachable countries. Further back, but still
within the top 10 were—alongside the USA and Great Britain—New Zealand in seventh
place (185 accessible countries) and Australia in eighth place (184 reachable travel destinations). This placed many states within the Asia-Pacific region APAC[6] high up on the
Passport Index (Tageskarte 2021).

As late as autumn 2020, many aviation experts believed that air travel would in future
require a Corona vaccination. A technological innovation, the so-called Travel Pass Initiative of the IATA, was supposed to contribute to the restoration of global mobility. In
the form of a mobile application (app), i.e. a digital platform, travellers were to be given
the opportunity to store and manage verified certification for Covid-19 tests or vaccinations (Tageskarte 2021; IATA 2021). But even the introduction of an electronic vaccine
passport proved to be more difficult than assumed and was delayed.

Price development is not an unimportant indicator for the current situation in tourism.
2020 was a significant cut, which also showed in the price development. However, tourism prices recovered already in 2021. In part, they even increased significantly compared
to the pre-Corona year 2019, as price comparisons in Switzerland showed: For example, fuel prices increased by 19.2% compared to the Corona year 2020 by mid-2021, but
were still 1.6% lower than in 2019. Overnight prices increased by 8.0% compared to the
previous year by mid-2021 and at least by 3.7% compared to 2019. Package holidays
cost 0.2% more in 2021 than in the first Corona year, but were still 12.7% lower than in
the pre-Corona period. Cable cars and ski lifts were −3.0% in 2021 compared to 2020
and −2.8% compared to 2019. Air fares fell by a further 4.5% in 2021 compared to 2020
and were 28.7% lower than in the pre-Corona year 2019 (Vontobel 2021, p. 14).

3.10   Tourism and the Economic Effects of Crises—An Endless Story?

Hall (2013, p. 23) has pointed out that tourism can be or can be affected by external crises, but also contributes to or exacerbates crises. So it is not without irony that tourism
has been seen as a possible solution to economic crises in countries like Greece, Spain
or Iceland. Thus, tourism proves to be an economic sector with high growth rates, high

---

[6]The Asia-Pacific region (APAC) includes a large part of Southeast Asia, Australia and Oceania.
Sometimes South Asia is also included, even though India and its neighboring countries do not
border the Pacific, but rather the Indian Ocean. Countries included in APAC also include Far East
Russia (North Pacific) as well as the countries of North and South America that lie on the east
coast of the Pacific Ocean (https://de.wikipedia.org/wiki/Asien-Pazifik).

Table 3.6 Cost structure of an independent travel agency. (From Cooper 2012, p. 2005 as well as Holloway and Taylor 2006)

Sales		1,500,000	
Gross profit (commission at average 9.4 %)		*141,000*	(100 %)
Expenditure			
Personnel			
Salaries, NHI, pensions	60,000		
Staff travel, training, subscriptions	*3000*		
		63,000	44.68 %
Establishment			
Rent, rates, water	22,000		
Light and heat	3000		
Insurance	2000		
Cleaning	*1300*		
		28,300	20.07 %
Administration			
Computers, telephone, website	10,000		
Postage	2000		
Printing & stationery	2000		
Hire of equipment	1000		
Advertising and publicity	4000		
Publications, timetables	*1000*		
		20,000	14.18 %
Financial and legal			
Credit cards	3500		
Bank charges	1500		
Auditing and accounting	4000		
Legal fees	500		
Bad depts	*500*		
		10,000	7.09 %
Depreciation and amortization		*4600*	3.26 %
Total operational costs		**125,900**	89.29 %
Net profit before tax		15,100	*10.71 %* 100 %
	Note: net profit as % of sales		1.10 %

foreign currency earnings and relatively high resilience to economic change. Conversely, tourism can cause considerable ecological damage in small economic areas due to excessive liberalization—and by no means always does the "pro-poor tourism policy"[7] achieve its declared goal of directly or indirectly improving the situation of poorer population groups.

The structure of travel providers and intermediaries has also changed as a result of increasing digitalization.

Table 3.6 shows the classical cost structure of an independent travel agency.

In contrast, digital booking platforms calculate differently. Above all, their infrastructure and personnel costs are lower, but the IT effort is significantly higher.

According to the view of Pillmayer and Scherle (2018, p. 15), there is hardly any industry that is as crisis-prone as tourism. Both authors point out that the possible causes of crises in tourism can cover a very wide range—ranging from terrorist attacks to natural disasters, service scandals, economic crises to local epidemics (Pillmayer and Scherle 2018, p. 15). However, there is still no reference to the fact that worldwide pandemics can have much more serious consequences than the immediate economic environment of the tourism industry, as the recent Covid-19 crisis has shown: border closures, stranded tourists, quarantine measures, refusal of home return of passengers and employees, e.g. on cruise ships, blocking of migrants and employees in the hospitality industry abroad, complete collapse of international tourism offers and international traffic. Of course, it is true that every crisis can also be an opportunity—this was shown, for example, by the strong domestic tourism in European countries during the Corona year 2020. But often other providers benefit from such changes than those affected by the failures.

The Corona pandemic basically only intensified a trend that had been apparent for 10 years: When up to 80% of travel agency sales threatened to collapse in the Corona year 2020 (Jordan and Vontobel 2020, p. 3) and a significant proportion of those employed in tourism, hospitality and catering became unemployed, the big question arose as to how this would affect the future development of tourism. Already in the years up to 2019, for example, six out of ten travel agencies had disappeared in Switzerland (Jordan and Vontobel 2020, p. 3), other countries knew similar developments.

References

Benz, Matthias 2020: Starker Sommer in den Schweizer Bergen. In: Neue Zürcher Zeitung vom 31.8.2020. 15.
Clancy, Michael 2011: Global Commodity Chains and Tourism. Past Research and Future Directions. In: Mosedale, Jan (Hrsg.): Political Economy of Tourism. A Critical Perspective. London/New York: Routledge. 75 ff.

[7] See the Chap. "4.9 Pro-Poor-Tourism".

Cook, Roy A./Hsu, Cathy H. C./Taylor, Lorraine L. 2018: Tourism. The Business of Hospitality and Travel. Sixth Edition. London/New York: Pearson.

Cooper, Chris 2012: Essentials of Tourism. Harlow: Pearson Education.

Descamps, Philipe 2020: Luftfahrt in Turbulenzen. In: Le Monde Diplomatique (deutsche Ausgabe Schweiz). Juli 2020. 10 f.

Dorsch, Monique 2016: Verkehr und Tourismus. Plauen: M&S-Verlag.

Elshaer, Abdallah M./Marzouk, Asmaa M. 2020: Labor in the Tourism and Hospitality Industry. Skills, Ethics, Issues, and Rights. Palm Bay/USA: Apple Academic Press AAP.

FIF 1995: Forschungsinstitut für Freizeit und Tourismus. Universität Bern: Moorschutz und Tourismus. Bern.

Frey, Bruno S. 2020: Venedig ist überall. Vom Übertourismus zum Neuen Original. Wiesbaden: Springer.

Freyer, Walter 2011: Tourismus. Einführung in die Fremdenverkehrsökonomie. 10., überarbeitete und aktualisierte Auflage. München: Oldenbourg Verlag.

Freyer, Walter 2015: Tourismus. Einführung in die Fremdenverkehrsökonomie. 11., überarbeitete und aktualisierte Auflage. Oldenbourg: De Gruyter.

Gross, Matilde S. 2017: Gesundheitstourismus. Konstanz: UK Verlagsgesellschaft.

Hall, C. Michael 2013: Financial Crises in Tourism and Beyond. Connecting Economic, Resource and Environmental Securities. In: Visser, Gustav/Ferreira, Sanette (Hrsg.): Tourism and Crisis. London/New York: Routledge. 12 ff.

Holloway, J.C./Taylor, N. 2006: The Business of Tourism. Harlow: Pearson Education.

Hopkins, Laura/Labonté, Ronald/Runnels, Vivien/Packer, Corinne 2018: Medical Tourism Today. What is the State of Existing Knowledge? In: Timothy, Dallen J. Hrsg.): Tourism Planning. Critical Concepts in Tourism. Volume IV: Contemporary Trends and Future Directions. London/New York: Routledge. 243 ff.

IATA 2021: IATA Travel Pass Initiative. https://www.iata.org/en/programs/passenger/travel-pass/ (Zugriff 15.3.2021).

Imwinkelried, Daniel 2020: Die Schweizer kommen – in diesem Jahr. In: Neue Zürcher Zeitung vom 13.6.2020. 10.

Jordan, Gabriela 2021: Luxus-Schiffe werben mit „coronafrei". In: Luzerner Zeitung vom 23.1.2021. 15.

Jordan, Gabriela/Vontobel, Niklaus 2020: Fast jeder Fünfte wird vor die Tür gestellt. In: Luzerner Zeitung vom 26.6.2020. 3.

Kaspar, Claude 1991: Die Tourismuslehre im Grundriss. 4., überarbeitete und ergänzte Auflage. St. Galler Beiträge zum Tourismus und zur Verkehrswirtschaft. Reihe Tourismus. Band 1. Bern/Stuttgart: Paul Haupt.

Kirstges, Torsten H. 2020: Tourismus in der Kritik. Klimaschädigender Overtourism statt sauberer Industrie? München: UVK Verlag.

Letzner, Volker 2014: Tourismusökonomie. Volkswirtschaftliche Aspekte rund ums Reisen. 2., überarbeitete und erweiterte Auflage. Oldenbourg: De Gruyter.

Lüthi, André 2020: „Diejenigen, die durchhalten, müssen die Preise erhöhen". Gespräch mit André Lüthi, CEO der Globetrotter Group, mit Susanna Müller. In: Neue Zürcher Zeitung vom 31.12.2020. 49.

Maxwell, Keely 2012: Tourism as Transaction. Commerce and Heritage on the Inca Trail. In: Lyon, Sarah/Wells, E. Christian (Hrsg.): Global Tourism. Cultural Heritage and Economic Encounters. Lanham/New York/Toronto/Plymouth: Altamira Press. 105 ff.

Mayer, Marius/Vogt, Luisa 2016: Bestimmungsfaktoren des Ausgabeverhaltens von Naturtouristen in den Alpen – das Fallbeispiel Simmental und Diemtigtal, Schweiz. In: Mayer, Marius/Job, Hubert (Hrsg.): Naturtourismus – Chancen und Herausforderungen. Studien zur Freizeit und Tourismusforschung. Band 12. Mannheim: Verlag Metagis-Systems. 99 ff.

Merkel, Janet 2018: Coworking: das Arbeitsmodell der Zukunft? In: Pechlaner, Harald/Innerhofer, Elisa (Hrsg.): Temporäre Konzepte. Coworking und Coliving als Perspektive für die Regionalentwicklung. Stuttgart: W. Kohlhammer. 33 ff.

Müller, Hansruedi 2011: Tourismuspolitik. Wege zu einer nachhaltigen Entwicklung. Glarus/Chur: Rüegger.

Müller, André 2021: Das Ökosystem Flughafen Zürich steckt weiter im Dauerstress. In: Neue Zürcher Zeitung vom 3.3.2021. 1.

Müller, Ute/Wysling, Andres/Pabst, Volker/Benz, Matthias 2021: Sind die Mittelmeerländer bereit? In: Neue Zürcher Zeitung vom 19.6.2021. 24 f.

Muntschick, Verena 2019: Überblick: Kontext, Kontrast, Konsequenzen. In: Zukunftsinstitut. Trendstudie: Der neue Resonanz-Tourismus. Herzlich willkommen. Frankfurt/Main: Zukunftsinstitut. 10 ff.

Nagos, Pythagoras 2020: „Die Kreuzfahrtgesellschaften werden sich kurzfristig einer Schadenskontrolle zuwenden". Interview mit Pythagoras Nagos, dem ehemaligen Managing Director von MSC Cruises in der Schweiz. Von Jean-Claude Raemy. In: Travelnews vom 28.5.2020. https://www.travelnews.ch/cruise/16100-die-kreuzfahrtgesellschaften-werden-sich-kurzfristig-einer-schadenskontrolle-zuwenden.html (Zugriff 25.4.2021).

Newsome, David/Moore, Susan A./Dowling, Ross K. 2013: Natural Area Tourism. Ecology, Impacts and Management. 2nd Edition. Bristol/Buffalo/Toronto: Channel View Publications.

Petrin, Susanna 2021: Dinieren bei Minusgraden. In: Neue Zürcher Zeitung vom 8.2.2021. 6.

Pillmayer, Markus/Scherle, Nicolai 2018: Krisen und Krisenmanagement im Tourismus. Eine konzeptionelle Einführung. In: Hahn, Silke/Neuss, Zeljka (Hrsg.): Krisenkommunikation in Tourismusorganisationen. Grundlagen, Praxis, Perspektiven. Wiesbaden: Springer VS. 3 ff.

Schmutz, Christoph/Müller, Ute/Wysling, Andres/Fischer, Thomas/Pabst, Volker/Belz, Nina 2020: Ferienorte versuchen den Sommer zu retten. In: Neue Zürcher Zeitung vom 20.5.2020. 18 f.

Schulz, Axel 2014: Modul B: Grundlagen Verkehr im Tourismus. Fluggesellschaften, Kreuzfahrten, Bahnen, Busse und Mietwagen. Kapitel 3: Kreuzfahrten. In: Schulz, Axel/Berg, Waldemar/Gardini, Marco A./Kirstges, Torsten/Eisenstein, Bernd: Grundlagen des Tourismus. Lehrbuch in 5 Modulen. 2., überarbeitete Auflage. München: Oldenbourg Verlag. 49 ff.

Statista 2021: Anzahl der Kreuzfahrtpassagiere aus Deutschland von 2004 bis 2020. https://de.statista.com/statistik/daten/studie/180388/umfrage/passagiere-von-kreuzfahrten-aus-deutschland/ (Zugriff 25.4.2021).

Strasdas, Wolfgang 2017: Einführung Nachhaltiger Tourismus. In: Rein, Hartmut/Strasdas, Wolfgang (Hrsg.): Nachhaltiger Tourismus. Einführung. 2., überarbeitete Auflage. Konstanz: UVK Verlagsgesellschaft. 13 ff.

Tageskarte 2021: 5. Januar 2021. https://www.tageskarte.io/tourismus/detail/asien-pazifik-region-dominiert-henley-passport-index-deutschland-auf-platz-3.html (Zugriff 24.6.2021).

Tietz, Bruno 1980: Handbuch der Tourismuswirtschaft. München: Verlag Moderne Industrie.

Vontobel, Niklaus 2021: Sommer der Inflation. In: Luzerner Zeitung vom 12.6.2021. 14.

Wattanakuljarus, Anan/Coxhead, Ian 2013: Is Tourism-based Development Good for the Poor? A General Equilibrium Analysis for Thailand. 182 ff. In: Dwyer, Larry/Seetaram, Neelu (Hrsg.): Recent Developments in the Economics of Tourism. Volume II. Tourism, Trade, Growth and Welfare. Cheltenham, UK/Northampton, MA: Edward Elgar. Ursprünglich in: Journal of Policy Modelling. 30 (2008). 929 ff.

Zenhäusern, Robert/Kadelbach, Thomas 2016: 12 Thesen zur Zukunft des Tourismus in den Berggebieten. Bern: Schweizer Tourismusverband/Schweizerische Arbeitsgemeinschaft für die Berggebiete. Juli 2018.

Zumbrunnen, Fabrice 2021: „Wir setzen vermehrt auf kleine Läden". Gespräch mit Migros-Chef Fabrice Zumbrunnen von Natalie Gratwohl und Andrea Martel. In: Neue Zürcher Zeitung vom 31.3.2021. 25.

Ecological Consequences of Tourism

4

The environment is a key resource for tourism. It consists of the socio-cultural and natural environment. Environmental damage caused by tourism not only reduces the quality of the natural environment, but also damages tourism. Mason (2017, p. 99) lists possible damage to the environment at tourist destinations, including erosion of footpaths, pollution of rivers, waterways and seas, waste, traffic jams, overpopulation and the creation of unsightly buildings. In addition, whole ecosystems can be endangered.

However, the destruction of the natural resources that are essential for tourism is unlikely to be stopped immediately. Rather, there is a danger of gradual, gradual destruction of the environment as well as the risk of environmental disasters (Rate et al. 2018, p. 7). And this development is likely to continue to some extent—until society develops and implements clear and binding tourism strategies.

According to Kirstges (2020, p. 59), tourism causes, among other things, additional land consumption and landscape sealing. So shielded infrastructure facilities are erected that can only be used by tourists, but not by locals. Private beaches of hotels or tourist resorts usually remain closed to locals, sometimes they also prevent access for local businesses, such as access to the sea for fishing. In many places, tourist ruins spoil the landscape. For example, the built-up area increased by 40% between 1987 and 2005 on Spain's coasts, and even by 60% on Spain's Mediterranean coasts. And by 2005, around a third of Spain's Mediterranean coast was built up in coastal areas (Kirstges 2020, p. 59).

The Turkish Riviera between Kemer, Antalya, Side and Anlanya is also largely built up, and in Bordum, the "St. Tropez of Turkey", almost 60% of the buildings had been erected illegally (Kirstges 2020, p. 69).

Since the 1980s, the negative aspects of tourism have also come into focus (Fuchs et al. 2017, p. 13). For example, tourism providers have been accused of exploiting the

C. J. Jäggi, *Tourism Before, During and After Corona*,
https://doi.org/10.1007/978-3-658-39182-9_4

local population or contributing to environmental destruction (Britton 1982 or Krippen-dorf 1986).

In his article on the relationship between environment and tourism, first published in 1992, Dowling (2015, p. 96) came to the conclusion that the relationship between environment and tourism is in a precarious balance, which can be better handled by a targeted integration of both, in order to better manage the existing interests and goal conflicts, which definitely exist. For this, a better coexistence in the sense of a mutual recognition of the respective interests is necessary.

But how great are the ecological impacts of tourism at all? The WWF has calculated an ecological footprint of the different types of holidays on the basis of the CO_2 emissions (Table 4.1).

Table 4.1 CO_2 emissions by type of holiday. *Sources* Dorsch (2016, p. 50) and WWF (2009, p. 9)

Vacation type (days)	Tourist climate footprint in CO_2 emissions per person in kg					
	Arrival and departure	Accommodation	Food	Activities on site	Total	Per day
Beach vacation Mallorca (14)	925	148	91	58	1221	87
	Airplane	Hotel	Full board	Car, motor-boat, squad		
Cultural holiday South Tyrol (5)	63	80	55	18	261	43
	Tour bus	Hotel	7 warm meals	Bus, ship, taxi		
All-inclusive vacation Mexico (14)	6361	487	205	165	7218	515
	Airplane	Hotel	25 hot meals	Flight, motorboat, ship		
Health vacation Allgäu (10)	105	110	73	5	297	29
	Train	Inn	17 hot meals	Rental car, cable car		
Ski vacation Vorarlberg (7)	296	85	32	10	422	60
	Pkw	Pension	11 warm meals	Pkw		
Mediterranean cruise (7)	685	439	79	21	1224	174
	Airplane	Cruise ship	11 warm meals	Excursions		
Vacation at home on the balcony (14)	0	17	9	33	59	4
	–	At home	Self-catering, restaurants	Car		

The question of how much tourism, tourist offers and tourists are acceptable and how much they are too big depends essentially on the carrying capacity *("carrying capacity")* of a place or region. According to Cook et al. (2018, p. 343), the carrying capacity is shown in four aspects or sub-criteria:

1. Physical capacity: number of tourists who can be accommodated in an area. This also includes the capacity of roads, parking spaces, available water resources and other properties of the location.
2. Environmental capacity: This is the number of visitors who can be received without reducing the attractiveness of the place. This depends on the subjective perception, the climatic conditions, the number of tourists and other aspects.
3. Ecological capacity: This is the number of tourists who can be received without damaging the ecology, such as without affecting the fauna and flora.
4. Social carrying capacity: This refers to the number of tourists that a society or a country can take in without causing substantial damage to the local culture.

If these sizes are exceeded, one can speak of overtourism *("overtourism")*.

In connection with questions of *overtourism*, Peeters et al. (2018, p. 26) have defined the following points as the desirable scope of tourism:

- Ecological capacity: only as much tourism as the environment can cope with without being overloaded.
- Physical conditions: Tourist projects and recreation activities must not impair the natural-physical environment.
- Socio-perceptive capacity: no tourism that leads to rejection by the population due to cultural or ecological impairments.
- Economic viability: Tourist activities must not marginalize, excessively compete with or destroy other economic activities.
- Psychological capacity: every *"overcrowding"* that the locals can no longer cope with is to be avoided.

As Pintassilgo and Silva (2013, pp. 367 ff.) have shown, the tourist accommodation industry depends not only on the quality of the environment, but also on the fact that the environment is not overused for tourism purposes. Unrestricted access not only leads to overuse of the natural environment, but also to economic exploitation and overuse of common goods. This affects the whole of tourism. At the same time, mass tourism attracts tourists with low willingness to pay, which already affects many tourist destinations today (Pintassilgo and Silva 2013, p. 380), further deteriorating the ecological situation of many tourist destinations. In other words: "Tourism can destroy tourism" (Pintassilgo and Silva 2013, p. 380). This vicious circle must be broken.

4.1   Mass and Cheap Tourism

With regard to mass tourism, Enzensberger (1962, p. 167) spoke of the "tristfulness" of the tourist. He wrote: "It is indeed very easy to make fun of the mass tourism of our days … But the power that today throws the masses to the beach of their little holiday happiness is tremendous all over the world. It is the power of a blind, unarticulated rebellion that constantly fails in the surf of its own dialectic. … The flood of tourism is a single escape movement from the reality with which our social system surrounds us. But every escape, however foolish, however powerless it may be, criticizes that from which it turns away".

In recent decades, travel has become progressively cheaper. For example, a flight from Zurich to Santiago de Chile in economy class cost an average of 1800 € in the 1970s. By early January 2021, the cheapest flight from Zurich to Santiago de Chile was 357 €.

High-speed trains have become cheaper in many places. Unlike the classic TGV, Ouigo trains served 42 destinations in France at the end of 2020 and connected the twelve largest cities in France by high-speed rail—and, according to Ouigo CEO Stéphane Rapebach, for no more than €25 for adults and at a flat rate for children (Belz 2020, p. 24). Although more passengers sit in Ouigo trains than in the TGV. Sockets have to be paid extra, tickets can only be booked online and at the latest 30 minutes before departure, the passengers have to scan their tickets at the platform and board the train. Before the Corona year 2020, the trains were 80–90% occupied, and with 18 million passengers, the Ouigo trains covered around 22% of high-speed traffic (Belz 2020, p. 24). It is interesting that the Ouigo customers did not simply jump off the more expensive TGV—so every second passenger did not drive by train before. Because the demand was also so great, the seats on routes such as Paris-Bordeaux were filled again after a temporary decline in the TGV.

However, the Corona pandemic also put a damper on train travel. Business travel in particular declined by 60% (Belz 2020, p. 24). With the foreseeable opening of high-speed rail to foreign companies, prices are certainly not expected to rise. But still 80% of the French made their journeys by private car (Belz 2020, p. 24).

Bauer (2017, p. 71) has pointed out that the "demand-driven production of something that would not exist without the customer", took place in parallel to modern tourism and at the same time increasingly raised the question of authenticity. The creation of a parallel world of accommodation, souvenirs, food, views and performances for tourists has not even spared the encounter with the stranger. Even large parts of the first ethnological collections in the nineteenth century—such as hooks, weapons, etc.—which served as great tourist magnets, turned out to be fakes afterwards. In particular, the development of tourist ghettos has had a wide impact with mass tourism. This conglomerate of tourism offers, services and organized encounters, which is referred to as the "tourist bubble", increasingly followed the established wishes of the audience in style and products

(Bauer 2017, p. 71). In essence, this artificial world—ranging from Disney parks to tourist accommodation, from designed natural beauties to folklore—has increasingly isolated tourists from everyday life on site and transported them into a world of illusion and parallelism that is interpreted by naive tourists as the reality of the country. This development is counterproductive and promotes intercultural misunderstandings and misperceptions.

This is particularly true of digital media. In fact, electronic media have changed the travel of dropouts and backpacker tourists to a great extent, as the following example shows (Kirstges 2020, p. 19):

"There are still dropout backpackers today. For example, Petra, a ... university graduate who takes a 'break' for long trips every year, reported from Cambodia in 2015: 'We even met a 87-year-old backpacker who spent two months in a simple hut there.' However, she also noticed[ed] her trip through Asia: What struck me on this trip is that you don't meet as many people as before, because most people have an iPhone and are busy with it. Before, you would have sat in a hostel somewhere and talked, now everyone sits by themselves and is typing on their iPhone, a bit sad, this development. Many people book everything in advance and are only on the road with a navigation system" (Kirstges 2020, p. 19; author's addition).

What a difference to travelling in the 1970s and early 1980s, when the author of these lines still travelled through Latin America alone, without a mobile phone and sometimes cut off from any possibility of telephonic communication for weeks—and if you were lucky, you would find a post office somewhere where you could call Europe for a few minutes against dozens of dollars, provided that the lines were not interrupted due to floods or bad weather. And finding accommodation could only be done on the spot …

Today, mobile applications lead to the creation of a virtual world, a predefined context for tourists, first of all. Secondly, the search for information is limited to these predefined fields of meaning—a self-confirming, self-referential tourism system arises. Thirdly, a comprehensive system of pre-bookings not only reduces or prevents possible spontaneity on the trip, the possible diversity of information is reduced and there is a "self-emancipation" (Bauer 2017, p. 83). Fourth, quality expectations are commercially standardized and "mass taste becomes the guideline" (Bauer 2017, p. 83). And fifth, this type of tourism "drives the individual tourists back to the herd they do not want to be" (Bauer 2017, p. 83).

4.2 Tourism and Climate Change

The largest impact on climate change in the tourism sector was and is air traffic. According to the WTO and the United Nations Environment Programme, in 2015 41% of the CO_2 emissions caused by tourism were due to air traffic, 34% to road and coach transport, 19% to accommodation and 2% to cruise ships (Mason 2017, p. 162). In 2015, tourism was responsible for at least 5% of global CO_2 emissions and for approximately 8% of global climate warming (Mason 2017, p. 162).

According to Descamps (2020, p. 10), 100 l of kerosene are burned per hour in current air traffic. While total greenhouse gas emissions have decreased by 19% in France since 1990, they have more than doubled in air traffic (Descamps 2020, p. 10). According to the International Civil Aviation Organization, a reduction in CO_2 emissions of 50% compared to 2005 is planned for 2050. However, if air traffic continues to increase at the same rate as up to 2019, emissions from aircraft would have to be around 90% lower than today—a goal that is simply not achievable with today's technology (Descamps 2020, p. 10).

No specific reduction targets for tourism have been agreed in previous climate agreements and agreements. However, tourist emissions reductions are partly the subject of other national reduction targets, such as in the areas of "buildings" or "transport" (Strasdas 2017, p. 56).

However, climate change does have an impact on tourism. Like no other economic sector, tourism is dependent on intact natural resources. For example, warmer, drier climate in southern latitudes is a key driver of tourist flows to these areas, such as from northern Europe to Mediterranean countries or from North America to Mexico or the Caribbean (Strasdas 2017, p. 57). In reverse, ski resorts depend on cold temperatures and snowfall. In addition to direct and indirect physical effects, climate change also has social consequences in travel behavior.

In winter tourism, there have already been major shifts in various places due to climate change. In France, many cross-country skiing areas saw a significant increase in guests during the winter of 2020/2021, while the ski slopes were closed—partly due to Corona, partly due to climate warming. In comparison to 2020, French winter resorts experienced a 47.8% drop in the spring of 2021, with an occupancy rate of only 33% based on booked beds (Descamps 2021, p. 18). It is interesting to note that the effects of climate warming vary depending on altitude: while people in middle altitudes thronged in masses and the ski resorts near the large cities of Grenoble, Chambéry and Annecy had as many day visitors as ever before—partly because in these places the ski lifts had not been running for years due to lack of snow and winter tourism had switched to alternative leisure activities at an early stage—the higher altitudes were most affected by the closure of the lifts and had up to 58% lower occupancy. Sports equipment manufacturers also benefited from the changed situation. For example, the French company TSL, the largest European manufacturer of snowshoes, sold 200,000 pairs of snowshoes in the winter of 2020/2021—that is one third of global sales—or 50,000 more than the year before (Descamps 2021, p. 18). The decline in downhill skiing has led and continues to lead to a greater social mix, the spread of other sports and leisure activities, but has also had negative consequences for the environment in many places (Descamps 2021, p. 18).

Overall, the warming of the earth was particularly evident in the mountain regions in the form of a particularly strong increase in temperature, especially in winter. Since 2014, this process has accelerated even further. For example, the winter of 2019/2020 was 3.3 degrees warmer than the average for the years 1961–1990 in the Alps. The amount of snow also fell by 24% (Descamps 2021, p. 19). Even if the lack of snow can be compensated for by artificial snow—"the environmental debt of alpine winter sports is growing year by year," says Descamps (2021, p. 19). The double effect of Corona and climate change should—according to Descamps (2021, p. 19)—be used to reduce alpine tourism—and to develop other, more gentle forms of tourism.

Tourism is also a contributing factor to climate change through its emissions. Conversely, tourist companies are also affected by climate change. Because, in particular, smaller tourist companies only plan for a period of 5–10 years, but the effects of climate change only become apparent after 20–50 years, this plays only a minor role for tourism at the moment—at least according to the opinion of Geigenmüller (2018, p. 109). However, it could be argued against this position that apparently climate-related environmental disasters have been increasing in recent years—in the form of floods, hurricanes, fires or the rise in sea level—which is why this assessment is likely to be too optimistic. Possible effects of climate change include interruptions in operations, seasonal and regional shifts in demand, and climate-related requirements for tourist infrastructure (Geigenmüller 2018, p. 110).

Thus, it can be said that tourism is on the one hand strongly dependent on climate and weather, but on the other hand also has a great influence on climate change itself. In principle—as with any other economic sector, only much stronger, as in agriculture—climate and weather form a basic requirement or a kind of natural resource for tourism, a kind of *conditio sine qua non,* that is, an indispensable condition. However, one must ask whether the tour operators are always aware of this: Although natural disasters are considered risks in tourism management, only a few tourism providers and actors do something proactively to prevent or slow down climate change. This is basically all the more surprising as climatic and meteorological developments have a direct impact on tourist investments (Scott et al. 2012, p. 75). It is also not less surprising that, in comparison to other macro- and microeconomic factors such as transport infrastructure, tourist demand markets, land ownership or access to the coast, weather and climate conditions are hardly ever mentioned as reasons for the tourist development[1] of new places or destination regions (Altalo and Hale 2020, p. XVI f. As well as Scott et al. 2012, p. 75).

Strasdas (2017, p. 77) has listed the possible adaptation measures to direct and indirect consequences of climate change for Germany (Table 4.2).

4.3   More Environmental Consequences of Tourism

A special environmental problem, especially in tourism, are liquid and solid waste as well as gaseous emissions.

---

[1] Or the abandonment of tourist development.

Table 4.2 Possible adaptation measures to direct and indirect consequences of climate change in Germany. (From Strasdas 2017, p. 77)

Risks/ Opportunities	Higher air temperatures	Lower precipitation	Extreme events	Impacts on biodiversity	Impacts on water bodies	Demand shifts
Possible adaptation strategies	Promotion of outdoor activities Use of the off-season	Promotion of Out-door activities Use of the off-season Water retention Water saving	Risk management Flood protection Floodplains	Resilience Strengthen ecosystems Alternate spaces/ biotope networks Create Dynamic nature conservation	Water retention/ renaturation Strengthen self-cleaning power Minimize use conflicts Ship/ engine technology Lock technology	New offers Quality management Foreign marketing Climate-friendly destinations Energy saving Environmental management Traffic planning

In the Mediterranean, tourism was responsible for 52% of all waste in 2017 (Réau and Guibert 2020, p. 3) and in the same year, the 47 ships of the Carival Corporation & PLC, to which Costa Croisières, P&O, Aida Croisières, Princess, Cunard Line, Seaborne and Holland-America Line belong, emitted ten times as much sulfur oxide as all 260 million cars in Europe, although this is not even a quarter of the European cruise ships (Réau and Guibert 2020, p. 3; Transport & Environment 2019, p. 6).

Even with so-called "nature tourism", the ecological footprint[2] is often very large—too often too large. This was also shown by the carbon footprint of safari tourism in Namibia about 10 years ago. For example, five tour operators investigated produced 4.9 to 6.8 t of greenhouse gases (CO_2) or 427–659 kg per travel day per guest, which at the time of the study was significantly above the global average for overseas travel of 0.5 t or 60 kg per guest and travel day. A trip to Namibia, for example, generates about 50% of the annual average consumption of CO_2 by a German citizen. Of the emissions, a total

---

[2] The so-called ecological footprint ("ecological footprint") is a method of determining the environmental impact of each individual. Based on approximately 5400 data points, it is calculated how much ecological capacity a country has (Wackernagel 2015, p. 66) and how much of it is used by humans. The basis of the footprint method is the biocapacity that is produced in one year. It should be noted that the calculation is only possible retrospectively (Wackernagel and Beyers 2010, p. 49). The footprint method focuses on the land and water areas that are necessary for the production of goods and services, but also for the disposal of waste (Wackernagel and Beyers 2010, p. 49). For a detailed discussion of the ecological footprint, see also Jäggi (2017, pp. 138 ff.).

of 84–90% fell on the outbound and return journey by plane (Strasdas and Zeppenfeld 2011, p. 71).

4.4 Tourism and Lifestyle

There is no doubt that tourism also has a great influence on the lifestyle of the local population. For example, an ethnographic study in Ameskar Fogani in Morocco showed that small-scale tourism brought financial benefits to the rural population, but also changed their lives. While people used to hardly let strangers into their houses, but had much more contact with other villagers, this has now reversed with the appearance of tourists. Strangers were now invited into the houses. One villager explained it as follows: "A lot has changed … Before, people invited each other for lunch, but today nobody lets anyone in anymore. Unless you need someone to work. Many things have changed. Before, people did everything together, today everyone has their own business. Before, everyone talked to each other, today everyone talks to themselves" (Metzger 2014, p. 56). However, one could object here that the change in village life was not only caused by the—numerically small—tourism, but above all a consequence of changed working and economic conditions. Tourism may be one cause of this, but it may also just be an expression of it.

Also on the part of the tourists the personal world view and the lifestyle decide, to what extent ecological travel offers are desired and booked. Figures 4.1 and 4.2 show the importance of the attitude to sustainability in the booking behavior of tourists in Germany and in Switzerland.

Aside from the fact that in Germany there are more "skeptical travelers" than in Switzerland, where there are more "balanced travelers" and slightly more ecologically oriented people, the distribution profiles are basically quite similar. It would be interesting

Fig. 4.1 Sustainability types in the booking behavior of tourists in Germany in 2011. (From Stettler and Wehrli 2013, p. 169; own representation)

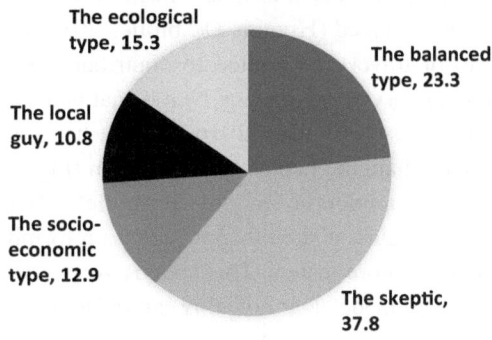

Fig. 4.2 Sustainability types in the booking behavior of tourists in Switzerland in 2011. *Source* Stettler and Wehrli (2013, p. 169); own representation

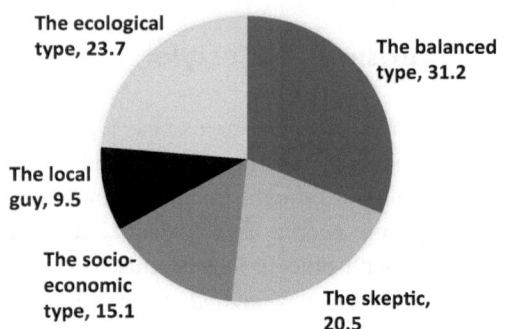

BOOKING BEHAVIOUR ACCORDING TO PERSONALITY TYPE IN SWITZERLAND 2011 IN %

to assign the different booking behavior to the respective voting behavior or the party preference, but unfortunately the corresponding data is missing.

Gössling (2015, p. 129) has also pointed out that tourism is also a key factor in the spread of diseases. For example, the probability that travelers from Europe or North America will become infected with diarrhea in southern countries is 30–80%, with hepatitis A 0.3%, with malaria 0.25% and with HIV 0.01%. The risk of infection with Covid-19 is likely to be significantly higher in many countries and regions for the foreseeable future than with hepatitis A or malaria.

4.5   Alternative Tourism—Soft Tourism as a Solution?

According to Higgins-Desbiolles (2016, p. 212), it is often forgotten today that the roots of alternative tourism lie in the 1960s. Understood as an alternative to mass tourism in times of high economic activity and its negative consequences, this form of tourism wanted to drive forward a kind of counterculture and be an alternative to the consumer society. So to speak, as a new form of development policy, international relations should also be changed (Higgins-Desbiolles 2016, p. 212). This made alternative tourism a new form of travel. He wanted to contribute to changing the world. This made alternative tourism very close to many NGOs that had a similar idea.

Higgins-Desbiolles (2016, p. 226) has schematically compiled the various forms of society-changing behavior and tourism (Fig. 4.3).

In the opinion of Higgins-Desbiolles (2016, p. 227), in contrast to globalized capitalism, just tourism should contribute to reforming societies and the world society and contribute to more justice. This sounds very nice—but can tourism actually do this? If so, it should be made clear how this can be done.

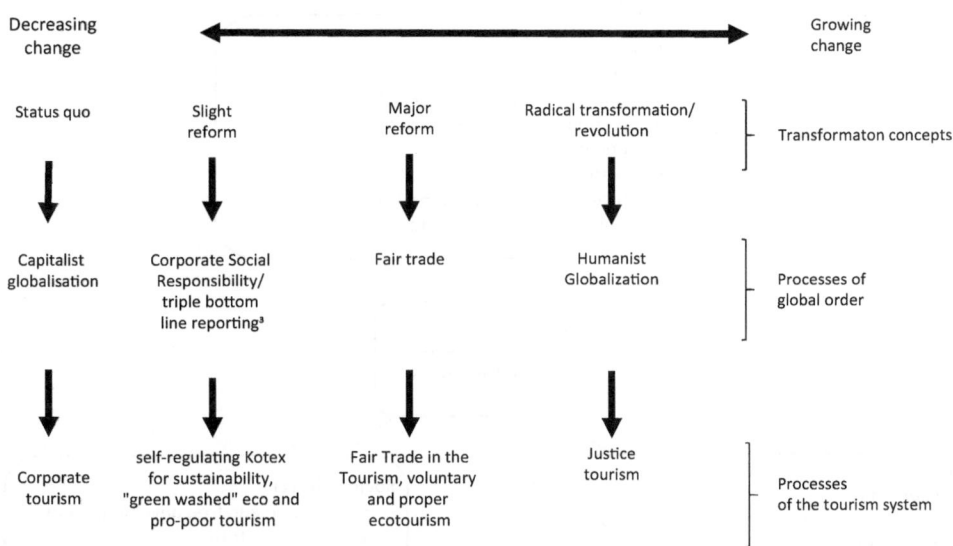

Fig. 4.3 Continuum of possible changes in the global world order and in tourism. [a]Triple Bottom Line is an accounting system that, in contrast to traditional accounting, not only calculates profit and loss, but also social and ecological aspects. (From Higgins-Desbiolles 2016, p. 226; Translation from English by the author)

Siegrist et al. (2015, p. 17) proposed using the following terms to describe tourism with the least possible interference in the natural environment, the least possible landscape consumption and the least possible changes to the landscape and to preserve a natural cultural landscape: agritourism, ecotourism, green tourism, ethical tourism, responsible tourism, fair tourism, geotourism and social tourism (Ständiges Sekretariat der Alpenkonvention 2013, pp. 15–16). But even here it applies: If the criteria and standards by which such forms of tourism are measured and evaluated are not clear, such terms are of no use whatsoever and remain "hot air".

Kirstges (2020, p. 142) has sketched a "magic triangle" for sustainable development and soft tourism (Fig. 4.4).

Siegrist et al. (2015, p. 18) have pointed out that the original demands of "soft tourism" would now be considered outdated because their demand has been replaced by a combination of the least possible wear of energy and natural resources with enjoyment and happiness. Rather, the approach of sufficiency is now in the center, which shows that sufficiency is possible with enjoyment, comfort and pleasure and without renunciation, as for example tourism in connection with successful architecture, energy or the Slow Food movement. As further examples, Siegrist et al. (2015, p. 18) mention e-bikes or exchange offices. Apart from the fact that all this is only more or less small niche areas—does anyone really believe that the massive effects of global mass tourism, such as the enormous kerosene emissions in air traffic or the combustion residues of tons of

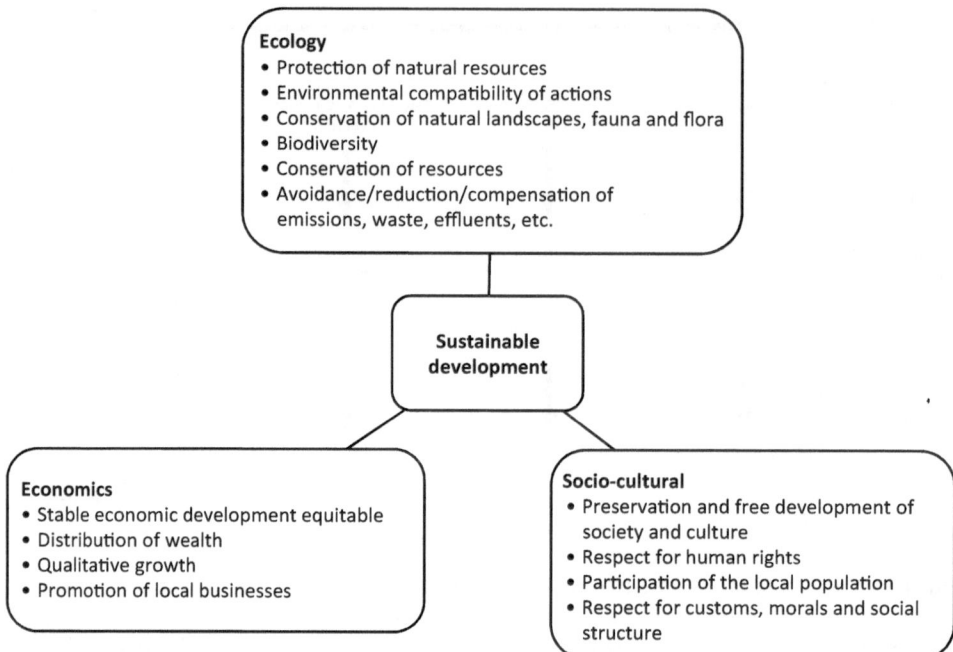

Fig. 4.4 "Magic triangle" of sustainable development and soft tourism. (Mod. after Kirstges 2020, p. 142; own representation and slightly edited by the author)

heavy oil by the cruise ships, can be eliminated by a few e-bikes or some Slow Food offers? Apart from the fact that e-bikes often burn nuclear power and millions of people can not afford alternative food or Slow Food at all. Against the background of the climate debate, it is now clear that an ecological transformation of society must necessarily also be accompanied by renunciation and less consumption.

4.6   Ecotourism

Fennell and Malloy (2016, p. 19) have suggested, following Ceballos-Lascurain (1991, p. 13), that ecotourism should be understood as tourism in unspoiled natural surroundings, carried out for the purpose of studying, admiring, and enjoying the local scenery, plants, animals, and any existing cultural manifestations (both past and present)[3]. However, this definition should be supplemented by four further aspects: first, ecotour-

---

[3] "Tourism in undisturbed natural areas with the specific objective of studying, admiring, and enjoying scenery, plants, and animals, as well as any existing cultural manifestations (both past and present) found in these areas" (Fennell and Malloy 2016, p. 19).

ism tries to have as little impact as possible on the social and ecological environment; second, ecotourism includes an ethic of behavior towards the natural environment; third, ecotourism can be understood as a method or instrument for the preservation of nature and natural surroundings; and fourth, ecotourism promotes employment and entrepreneurial activity among the local population (2016, p. 19).

Wearing et al. (2012, p. 39) estimated that ecotourism accounts for 3–4% of global tourism, depending on the definition of ecotourism used.

According to Siegrist et al. (2015, p. 19), ecotourism originally emerged from long-distance tourism in large protected areas in Africa, Asia and Latin America. Thus, non-governmental organizations wanted to establish ecotourism as a special form of sustainable tourism together with the World Tourism Organization UNWTO after the Rio Conference on the Environment in 1992. With ecotourism as an environmentally and socially compatible form of tourism, in particular in less developed areas and in emerging countries, the aim was to combine travel to intact natural areas with the preservation of biodiversity and natural resources, while at the same time ensuring sufficient value creation, which makes the financing of protected areas possible, creates income opportunities for the local population and makes them aware of the need to protect nature. Partly criticized was the often lacking sustainability in practice and the exclusive focus on protected areas (Siegrist et al. 2015, p. 19). Because the conditions for tourism in the Alps were different—for example as a historical consequence of industrialization—ecotourism is only very limited suitable for these regions.

Figure 4.5 shows the different ecological and sustainable concepts of tourism together.

According to Siegrist et al. (2015, p. 21), however, the types of tourism listed in Fig. 4.5 are not always clearly separated. For example, both nature-based tourism and ecotourism are strongly nature-oriented, but ecotourism is not necessarily sustainable (e.g. for long-distance travel). Conversely, sustainable tourism does not necessarily have a nature orientation, because sustainable tourism can also be city tourism. Rural tourism and agrotourism do not necessarily have to be nature-oriented or sustainable in the same way (Siegrist et al. 2015, p. 21).

According to Grimm et al. (2012, p. 32), in 2010 42% of the German population said they would like to go on a holiday to the countryside at some point, 24% had already done so. The more specific overall interest [4] in Germany in taking a holiday to the countryside was between 14% and 10% of those surveyed between 2000 and 2011, with a slight downward trend. More than half of those interested in a holiday to the countryside lived in a household with children (Grimm et al. 2012, p. 33).

In a survey of visitors to Northeast Brandenburg in the winter semester of 2005/2006, 84% described an attractive landscape as very important and a further 11% as important. The visitors considered the animals on site, horse riding, hiking, cycling and cultural

---

[4] i.e. the declared intention to go on a holiday to the countryside in the next 3 years.

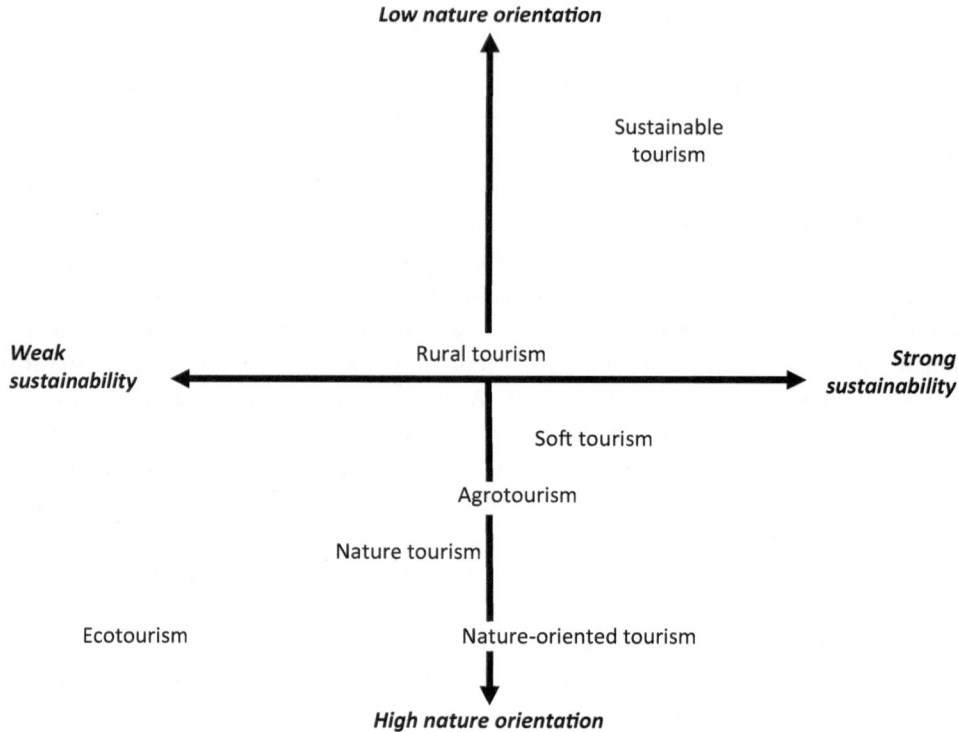

Fig. 4.5 Different orientation of different ecological and sustainable tourism concepts. (From Siegrist et al. 2015, p. 20; Siegrist and Ketterer Bonnelame 2016, p. 48; own representation)

attractions to be much less important (Peters 2012, p. 61). This probably indicates that, at least in rural tourism, nature and the landscape are of decisive importance—and thus the most important resource for tourism.

A survey of 1365 people working in the tourism industry, research and NGOs, and stakeholders of tourism products carried out in 2012 found that the proportion of nature-based tourism in the Alps was estimated at around 15–25%. 18.8% of those surveyed believed that the economic development of the Alps was more important than the protection of the Alps (Siegrist et al. 2015, p. 151). Not surprisingly, this group of more economically oriented people wanted no or only few state regulations. The protection of nature reserves was not the focus for this group—in contrast to other respondents—nor did they support a financial contribution to the protection of natural resources and protected areas.

Cater (2015, pp. 271 ff.) has criticized ecotourism as a "Western construct", as an expression of cultural hegemony and as a "form of paternalism". In particular, the diverging values and interests between tourism providers and representatives of ecotourism would often be overlooked (Cater 2015, p. 281). For example, the view of nature, but also the understanding of development, would be the result of complex social interactions of the most diverse social groups—and that could not simply be covered up by well-meaning Western academic concepts or considerations. Rather, the local communities would have to tell their own, authentic stories and bring in and implement their own ideas (Cater 2015, p. 285).

4.7   Away from Gigantism—But Not All Have Noticed it yet

Shortly after the Klein Matterhorn in Zermatt was made accessible by a new cable car, which could transport not 600 but 2000 tourists per hour instead of the previous one, it became known that Zermatt was planning another gigantic tourism project: an "Alpine crossing" or "Alpine crossing" by cable car from Italy over the Alps to Zermatt (Wirth 2018, p. 3). For this purpose, a cable car gap from the Klein Matterhorn to the Italian side of the Alps should be closed and led to the Testa Grigia mountain station, with a cost of 25 million francs. This would give a cable car connection from the Italian Cervinia in the Aosta Valley to Zermatt over the Alps. The tourism director raved about a completely new feeling that the Alpine crossing with the cable car, which should lead up to almost 4000 m above sea level, and thousands of new tourists from Asia—and he compared the project with the Alpine crossing of Hannibal (Wirth 2018, p. 3). This should create an additional value added of 700 million francs per year, with up to 35,000 tourists staying in Zermatt in the high season in winter 2018. Only one objection, namely that of Raimund Rodewald from the Stiftung Landschaftsschutz, was still outstanding at the end of 2018. He referred, among other things, to the fact that parts of the railway would not be located in the tourist use zone, he emphasized the need for nature conservation in the high mountains and warned of the "Disneylandization of the Alpine region" (Wirth 2018, p. 3). But according to the assessment of many observers, the railway would definitely come—the opening was planned for autumn 2021.

In Austria, too, opposition to large tourist infrastructure projects has been growing in recent years, such as in the Stubaital valley near Innsbruck, where a new railway and additional pistes are to be built in a difficult-to-access area (Imwinkelried 2021, p. 21). The additional value of such projects is disputed—and more and more people are against them for reasons of landscape conservation. However, many tourism experts continue to plan cheerfully. For example, railway officials in Sölden and the Pitztal in Tyrol are planning to connect several facilities across a glacier (Imwinkelried 2021, p. 21). Despite climate change, a ski area connection is planned in Hinterstoder, even though the ski lifts are only at an altitude between 800 and 1100 m—too low for a winter-proof skiing offer.

4.8   Community Based Tourism

The Community Based Tourism postulates local control over tourism and a win-win situation for rural communities (Blackstock 2015, p. 61). This approach requires the increased involvement of the local community in tourism and in the design of tourism offers. But in practice this raises questions (Blackstock 2015, pp. 63 ff.): Firstly, the local community is involved, but it does not have the opportunity to reject tourism as a development option if necessary. Secondly, the question arises as to who can legitimately speak for the community—most communities are very heterogeneous and the various actors often represent different interests. Thirdly, there are overriding or external barriers that prevent a community from actually participating in or even controlling tourism: for example, the international competition between destinations, economic laws and costs incurred outside the local community.

4.9   Pro-Poor-Tourism

On another level of the postulate of sustainable tourism lies the concept of pro-poor tourism (PPT), that is, tourism for the poor. PPT is less an economic niche, but rather an orientation of tourism (Weeden 2014, p. 9).

For tourism oriented towards the poor, three core requirements are important: First, the access of the poor to the economic benefits of tourism should be increased, for example in terms of business prospects, employment opportunities and income. This requires training and education so that the poor can take advantage of opportunities to increase their income. Secondly, the negative aspects and impacts of tourism in social and environmental terms must be minimized—such as restricted access to land, coastal areas or other natural resources, as well as protection against social disadvantage and exploitation. Thirdly, there is a need for a tourism policy and tourist planning that guarantees the participation of the poor in planning and decision-making, promotes partnerships between the private tourism sector and the poor, and develops new tourist products (Weeden 2014, p. 9). Pro-poor tourism usually postulates connections between tourism offerings and local informal service providers. According to Mundt (2011, p. 73), the big problem for local informal providers is that they are basically disadvantaged in order to formalize contracts with national or international partners. Without access to public goods such as the institutions of the public legal system, it is, according to Mundt (2011, p. 73), almost impossible for local, informal business partners to successfully escape poverty.

Perhaps this is why the efforts of pro-poor tourism strategies have focused more on opening up employment and income opportunities for the poor within the framework of tourist offers, rather than on bringing tourism to other areas or sectors, thus making the cake more accessible than expanding it[5]. Not entirely without controversy, Mundt (2011, p. 74) concludes that pro-poor tourism is in danger of becoming a purely marketing instrument of development organizations and NGOs if it is not clarified beforehand what is meant by sustainability in tourism.

---

[5] "'tilting' not expanding the cake" (Mundt 2011, p. 73).

4.10 Corona as the Cause of the Ecological Recovery of Tourist Destinations

At many tourist destinations, the Corona break led to a real ecological recovery of the environment. For example, the mayor of Maspalomas-Costa Canaria, Maria Concepción Narváez Vega, with its famous dunes and two sandy beaches, which could hardly save itself from tourists from Northern Europe before the Corona crisis, explained: "During the long lockdown, nature … recovered, the sand dunes reformed, we cleaned everything, and now we see animals again that we had hardly ever seen before" (Müller 2020, p. 20). In contrast to summer 2019, when 500 planes with guests from abroad arrived on the Canary Islands every day, the number of arriving tourists shrank by 80% in 2020. The few guests benefited from the paradise-like peace—but only every fourth hotel bed was occupied (Müller 2020, p. 20). Accordingly, the responsible persons on the Canary Islands are trying to get away from the tourist monoculture and settle other industries. Traditional agriculture is to be promoted, for example, and new technologies are to be settled, but also the film industry—among other things by means of a very low tax rate of 4% (Müller 2020, p. 20).

4.11 Increasing Tourism or Travel Costs?

Three things can be said with certainty about the future development of tourism: First, the number of people who want to travel and use tourist services will increase in the medium and long term—at the latest when the new middle classes in the populous countries push onto the tourism market. This development was already clearly visible in Chinese, Indian and Pakistani tourists before Corona. Secondly, the number of tourist destinations and their capacity to receive tourists will not increase indefinitely—at some places they have already reached or even exceeded the ceiling. This means thirdly that in the medium and long term, either tourism contingents will have to be set up in the countries of entry or travel will become very expensive if the price is used for quantity control. Possibly both will happen.

Not only at the destinations, but also during transport, ecological requirements are likely to make travel more expensive, for example if—in contrast to the current situation in many countries—additional and possibly massive environmental taxes are levied on air and road traffic. The former is likely to affect long-distance travel in particular, the latter internal and local tourism.

A global tourism catalogue will probably prove to be essential for its length, the partly anarchic world tourism is likely to increase significantly in planning and control. This also increases tourism costs in the long term.

4.12　Couchsurfing as a Cheap Accommodation Alternative?

Couchsurfing has established itself as a cheap and uncomplicated way of accommodation, especially among young people. The accommodation is free. Couchsurfers can be both guests and hosts—the minimum age is 18 years. The participants create a profile on Facebook with a description of their own person and present the sleeping place. The stay

Table 4.3 Advantages and disadvantages of couch surfing from the perspective of the guest. (Mod. after Hartmann and Pasel 2014, p. 98; own representation and slightly edited by the author)

Couchsurfing	
Advantages	Disadvantages
"Participation" possible for everyone (from 18 years on)	Registration required Creating a user profile Disclosure of personal data to community Success depends on "subcultural capital"
Free accommodation	Obligation of reciprocity Expectations of the host Short stay
Accommodation in private space Insights into local lifestyle + accommodation Relative independence/freedom (e.g. use of kitchen) Feeling of belonging + home	No special facilities for disabled, seniors, children No or limited privacy Strongly limited "comfort" Conflict avoidance/resolution only possible by leaving the accommodation Rules of the host (e.g. night rest): Compromises necessary
Temporal flexibility (arrival/departure/stay/last minute)	Only after registration Search for "suitable" host takes time
Unobligation: Cancellation by guest possible	Unobligation: Cancellation by host before/upon arrival or during stay possible
Selection of the host online before stay Recommendations by like-minded people	Host looks for "exotics" Predominantly young members Security risks
Host multilingual Host open to other cultures	Conflict potential: different views of openness Stereotypes, prejudices
("Intimate") interactions with host Intercultural exchange Activities with host (outside the tourism landscape)	Conflict potential e.g. due to different concepts of hospitality "bligation" to spend time with host
Individual and unplanned travel experiences, subjectively experienced as authentic	
Ability to find new friends/short-term friends, build networks and exchange beyond stay	

should always be short—one to three nights. Cancellations are possible and references are visible to everyone.

The average age of couch surfers is 28 years and the gender ratio is relatively balanced (Hartmann and Pasel 2014, p. 97).

Couchsurfing.org has been online since 2004 (Hartmann and Pasel 2014, p. 93). At the center of the idea are cosmopolitanism, openness, community, hospitality without market principles or monetary interests, mutual exchange, online planning and mutual acceptance of private space (Hartmann and Pasel 2014, pp. 94 ff.). In 2014, 7 million members were registered in more than 100,000 cities around the world (Hartmann and Pasel 2014, p. 93).

Table 4.3 shows the advantages and disadvantages of couch surfing from the perspective of the guest.

In contrast to other providers such as Airbnb, Couchsurfing is free and not commercially oriented. Because accommodation is moving within the framework of personal contacts and friendship—where there is currently an empty couch—there are also fewer problems with landlords than with Airbnb.

References

Altalo, Mary G./Hale, Monica 2002: Requirements of the US Recreation and Tourism Industry for Climate, Weather and Ocean Information. Consultant's Report to National Oceanic and Atmospheric Administration. May 2002. https://www.yumpu.com/en/document/read/4621741/requirements-of-the-us-recreation-and-tourism-industry-for-gcoos (Zugriff 12.3.2021).

Bauer, Ulrich 2017: Happy Idiots inside the M-Bubble. Ethische und soziale Implikationen der Abhängigkeit von elektronischen Hilfsmitteln beim Reisen. In: Landvogt, Markus/Brysch, Armin A./Gardini, Marco A. (Hrsg.): Tourismus – E-Tourismus – M-Tourismus. Herausforderungen und Trends der Digitalisierung im Tourismus. Berlin: Erich Schmidt Verlag. 71 ff.

Belz, Nina 2020: Die Billigbahn Ouigo bringt Franzosen in den Zug. In: Neue Zürcher Zeitung vom 26.11.2020. 24.

Blackstock, Kirsty 2015: A Critical Look at Community Based Tourism. In: Sharpley, Richard (Hrsg.): Tourism and Development. Volume III. Tourism Alternatives. Los Angeles/London: Sage. 61 ff. Ursprünglich in: Community Development Journal. 40/1 (2005). 39 ff.

Britton, S. 1982: The Political Economy of Tourism in the Third World. In: Annals of Tourism Research. 9/3 (1982). 331 ff.

Cater, Erlet 2015: Ecotourism as a Western Construct. In: Sharpley, Richard (Hrsg.): Tourism and Development. Volume II. Tourism and Sustainable Development. Los Angeles/London: Sage. 271 ff. Ursprünglich in: Journal of Ecotourism. 5/1&2 (2006). 23 ff.

Ceballos-Lascurain, H. 1991: The Future of Ecotourism. In: Lindberg, K. (Hrsg.): Policies for Maximizing Nature Tourism's Ecological and Economic Benefits. Washington, DC: World Resources Institute. 1 ff.

Cook, Roy A./Hsu, Cathy H. C./Taylor, Lorraine L. 2018: Tourism. The Business of Hospitality and Travel. Sixth Edition. London/New York: Pearson.

Descamps, Philipe 2020: Luftfahrt in Turbulenzen. In: Le Monde Diplomatique (deutsche Ausgabe Schweiz). Juli 2020. 10 f.

Descamps, Philipe 2021: Après Ski. Die Berge emanzipieren sich vom Pistensport. In: Le Monde Diplomatique (deutsche Ausgabe Schweiz). April 2021. 1/18/19.

Dorsch, Monique 2016: Verkehr und Tourismus. Plauen: M&S-Verlag.

Dowling, R. K. 2015: Tourism and Environmental Integration: The Journey from Idealism to Realism. In: Sharpley, Richard (Hrsg.): Tourism and Development. Volume I. The Tourism-Development Dilemma: The Benefits and Costs of Tourism. Los Angeles/London: Sage. 87 ff. Ursprünglich in: Cooper, C. P./Lockwood, A. (Hrsg.): Progress in Tourism, Recreation and Hospitality Management. Volume 4. London 1992: Belhaven Press.

Enzensberger, Hans Magnus 1962: Eine Theorie des Tourismus. In: Enzensberger, Hans Magnus: Einzelheiten. Frankfurt/Main: Suhrkamp. 147 ff.

Fennell, David A./Malloy, David Cruise 2016: Ethics and Ecotourism. A Comprehensive Ethical Model. In: Fennell, David (Hrsg.): Tourism Ethics. Critical Concepts in Tourism. Volume III: Types of Tourism an Ethics. London/New York: Routledge. 18 ff.

Fuchs, Matthias/Abadzhiev, Andrey/Svensson, Bo/Höpken, Wolfram 2017: A Knowledge-Based Paradigm for the Governance of Destination Sustainability. In: Pechlaner, Harald/Keller, Peter/Pichler, Sabine/Weiermair, Klaus (Hrsg.): Changing Paradigms in Sustainable Mountain Tourism Research. Problems and Perspectives. International Tourism Research and Concepts. Volume 7. Berlin: Erich Schmidt Verlag. 13 ff.

Geigenmüller, Michelle 2018: Qualifizierung von Tourismusunternehmen zur Entwicklung innovativer Klimaanpassungskonzepte. In: Mosedale, Jan/Voll, Frieder (Hrsg.): Nachhaltigkeit und Tourismus – 25 Jahre nach Rio – und jetzt? Mannheim: Verlag Metagis-Systems. 109 ff.

Gössling, Stefan 2015: Global Environmental Consequences of Tourism. In: Sharpley, Richard (Hrsg.): Tourism and Development. Volume I. The Tourism-Development Dilemma: The Benefits and Costs of Tourism. Los Angeles/London: Sage. 101 ff. Ursprünglich in: Global Environmental Change. 12/4 (2002). 283 ff.

Grimm, Bente/Schmücker, Dirk/Ziesemer, Kai 2012: Nachfrage und Kundenpotenziale für den ländlichen Tourismus. In: Rein, Hartmut/Schuler, Alexander (Hrsg.): Tourismus im ländlichen Raum. Wiesbaden: Springer Gabler. 27 ff.

Hartmann, Rainer/Pasel, Sandra 2014: Couchsurfing als innovative Übernachtungsalternative. Entwicklung und Prognosen als Herausforderung für gewerbliche Beherbergungsbetriebe. In: Küblböck, Stefan/Thiele, Franziska (Hrsg.): Tourismus und Innovation. Studien zur Freizeit- und Tourismusforschung. Band 10. Mannheim: Metagis-Systems. 89 z.

Higgins-Desbiolles, Freya 2016: Justice Tourism and Alternative Globalisation. In: Fennell, David (Hrsg.): Tourism Ethics. Critical Concepts in Tourism. Volume I: Theories of Ethics and Tourism. London/New York: Routledge. 210 ff. Ursprünglich in: Journal of Sustainable Tourism. 16/3 (2008). 345 ff.

Imwinkelried, Daniel 2021: Die Tourismus-Lobby stösst auf Opposition. In: Neue Zürcher Zeitung vom 19.4.2021. 21.

Jäggi, Christian J. 2017: Ökologische Baustellen aus Sicht der Ökonomie. Verlierer – Gewinner – Alternativen. Wiesbaden: Springer Gabler.

Kirstges, Torsten H. 2020: Tourismus in der Kritik. Klimaschädigender Overtourism statt sauberer Industrie? München: UVK Verlag.

Krippendorf, J. 1986: Tourism in the System of Industrial Society. In: Annals of Tourism Research. 13/4 (1986). 517 ff.

Mason, Peter 2017: Geography of Tourism. Image, Impact and Issues. Oxford: Goodfellow Publishers.

Metzger, J. 2014: „Arbeit ist nur das, was Geld bringt". Wandel der lokalen Ökonomie in Ameskar Fogani (Marokko) am Beispiel des Tourismus. In: Geographica Helvetica. Nr. 69 (2014). 49 ff.

Müller, Ute 2020: Die Kanarischen Inseln suchen Alternativen zur touristischen Monokultur. In: Neue Zürcher Zeitung vom 16.7.2020. 20.

Mundt, Jörn W. 2011: Tourism and Sustainable Development. Reconsidering a Concept of Vague Policies. Berlin: Erich Schmidt Verlag.

Peeters, P., Gössling, S., Klijs, J., Milano, C., Novelli, M., Dijkmans, C., Eijgelaar, E., Hartman, S., Heslinga, J., Isaac, R., Mitas, O., Moretti, S., Nawijn, J., Papp, B. and Postma, A. 2018: Overtourism. Impact and Possible Policy Responses. Research for TRAN Committee. Brussels: European Parliament. Policy Department for Structural and Cohesion Policies.

Peters, Jürgen 2012: Ortsbild und Landschaftsstruktur als Grundlage des ländlichen Tourismus. In: Rein, Hartmut/Schuler, Alexander (Hrsg.): Tourismus im ländlichen Raum. Wiesbaden: Springer Gabler. 45 ff.

Pintassilgo, Pedro/Silva, João Albino 2013: „Tragedy of the Commons" in the Tourism Accommodation Industry. In: Dwyer, Larry/Seetaram, Neelu (Hrsg.): Recent Developments in the Economics of Tourism. Volume I. Demand, Supply, Pricing, Taxation, Employment and the Environment. Cheltenham, UK/Northampton, MA: Edward Elgar. 367 ff. Ursprünglich in: Tourism Economics. 12/2 (2007). 209 ff.

Rate, Shirley/Moutinho, Luiz/Ballantyne, Ronnie 2018: The New Business. Environment and Trends in Tourism. In: Moutinho, Luiz/Vargas-Sánchez, Alonso (Hrsg.): Strategic Management in Tourism. 3rd Edition. Oxfordshire/Boston: Cabi. 1 ff.

Réau, Betrand/Guibert, Christophe 2020: Wie geht guter Tourismus? In: Le Monde Diplomatique (deutschsprachige Ausgabe Schweiz). Juli 2020. 3.

Scott, Daniel/Halland, C. Michael/Gössling, Stefan 2012: Tourism and Climate Change. Impacts, Adaptation and Mitigation. London/New York: Routledge.

Siegrist, Dominik/Ketterer Bonnelame, Lea 2016: Qualitätsstandards für den naturnahen Tourismus in den Alpen. In: Bieger, Thomas/Beritelli, Pietro/Laesser, Christian (Hrsg.): Gesellschaftlicher Wandel als Herausforderung im alpinen Tourismus. Schweizer Jahrbuch für Tourismus 2015/2016. Berlin: Erich Schmidt Verlag. 47 ff.

Siegrist, Dominik/Gessner, Susanne/Ketterer Bonnelame, Lea 2015: Naturnaher Tourismus. Qualitätsstandards für sanftes Reisen in den Alpen. Bern: Haupt Verlag.

Ständiges Sekretariat der Alpenkonvention 2013: Nachhaltiger Tourismus in den Alpen. Alpenzustandsbericht 4. Alpensignale – Sonderserie 4. Innsbruck: Ständiges Sekretariat der Alpenkonvention.

Stettler, Jürg/Wehrli, Roger 2013: Das Verständnis von nachhaltigem Tourismus in der Schweiz und in Deutschland. In: Bieger Thomas/Beritelli, Pietro/Laesser, Christian (Hrsg.): Nachhaltigkeit im alpinen Tourismus. Schweizer Jahrbuch für Tourismus 2012. Berlin: Erich Schmidt Verlag. 159 ff.

Strasdas, Wolfgang 2017: Herausforderungen an den nachhaltigen Tourismus. In: Rein, Hartmut/Strasdas, Wolfgang (Hrsg.): Nachhaltiger Tourismus. Einführung. 2., überarbeitete Auflage. Konstanz: UVK Verlagsgesellschaft. 45 ff.

Strasdas, Wolfgang/Zeppenfeld, Runa 2011: Naturtourismus und Ökotourismus. In: Antz, Christian/Eisenstein, Bernd/Eilzer, Christian (Hrsg.): Slow Tourism. Reisen zwischen Langsamkeit und Sinnlichkeit. München: Martin Meidenbauer. 55 ff.

Transport & Environment 2019: European Federation for Transport an Environment: One Corporation to Pollute them All. Luxury Cruse Air Emissions in Europe. Brussels: Juni 2013. https://www.transportenvironment.org/sites/te/files/publications/One%20Corporation%20to%20Pollute%20Them%20All_English.pdf (Zugriff 27.2.2021).

Wackernagel, Mathis 2015: Die ideale Welt ist keine Footprint-Welt – Interview mit Mathis Wackernagel. In: Beyers, Bert et al. (Hrsg.): Grosser Fuss auf kleiner Erde? Bilanzieren mit dem Ecological Footprint. Anregungen für eine Welt begrenzter Ressourcen. 64 ff.

Wackernagel, Mathis/Beyers, Bert 2010: Der Ecological Footprint. Die Welt neu vermessen. Hamburg: Europäische Verlagsanstalt.

Wearing, Stephen/Wearing, Michael/McDonald, Matthew 2012: Slow'n Down the Town to Let Nature Grow: Ecotourism, Social Justice and Sustainability. In: Fullagar, Simone/Markwell, Kevin/Wilson, Erica (Hrsg.): Slow Tourism. Experiences and Mobilities. Bristol/Buffalo/Toronto: Channel View Publications. 36 ff.

Weeden, Clare 2014: Responsible Tourist Behaviour. London/New York: Routledge.

Wirth, Dominic 2018: Kampf um die Berge. In: Luzerner Zeitung vom 23.11.2018. 3.

WWF 2009: Der touristische Klima-Fussabdruck. Hamburg: Internationales WWF-Zentrum für Meeresschutz.

Tourism and Infrastructure

Tourism offers are strongly linked to local infrastructure, but also to local, national and international transport infrastructure. For some types of tourism—such as event tourism—this is more the case than for others.

5.1 Event Tourism

Müller (2011, p. 61) has compiled various types of events based on Freyer (2000, p. 352) (Fig. 5.1).

Zenhäusern and Kadelbach (2018, p. 27) have pointed out that investments for the core infrastructure in mountain tourism are very capital intensive. This is also true for other areas of tourism, such as mega-events. In many cases, public funds are used in such cases. However, many of these investments are made without overarching development plans, which often does not guarantee their sustainability (Zenhäusern and Kadelbach 2018, p. 27).

In an economic study on the effects and promotion-worthiness of tourist events in Switzerland, Stettler et al. (2016, p. 277) came to the conclusion that smaller events with an international reach have a better cost-benefit ratio than large or mega-events. The authors called for each event to be examined in terms of target strategy in order to achieve the greatest possible match between target group and core offering and to optimally use the media effect of the event. Only in this way could events be used effectively as instruments for economic and tourism promotion. The involvement of the population would also be necessary to successfully anchor an event in the region. However, the effects of many events would be analyzed during and after the event only insufficiently from the perspective of tourism promotion (Stettler et al. 2016, p. 277).

© The Author(s), under exclusive license to Springer Fachmedien Wiesbaden GmbH, part of Springer Nature 2022
C. J. Jäggi, *Tourism Before, During and After Corona*,
https://doi.org/10.1007/978-3-658-39182-9_5

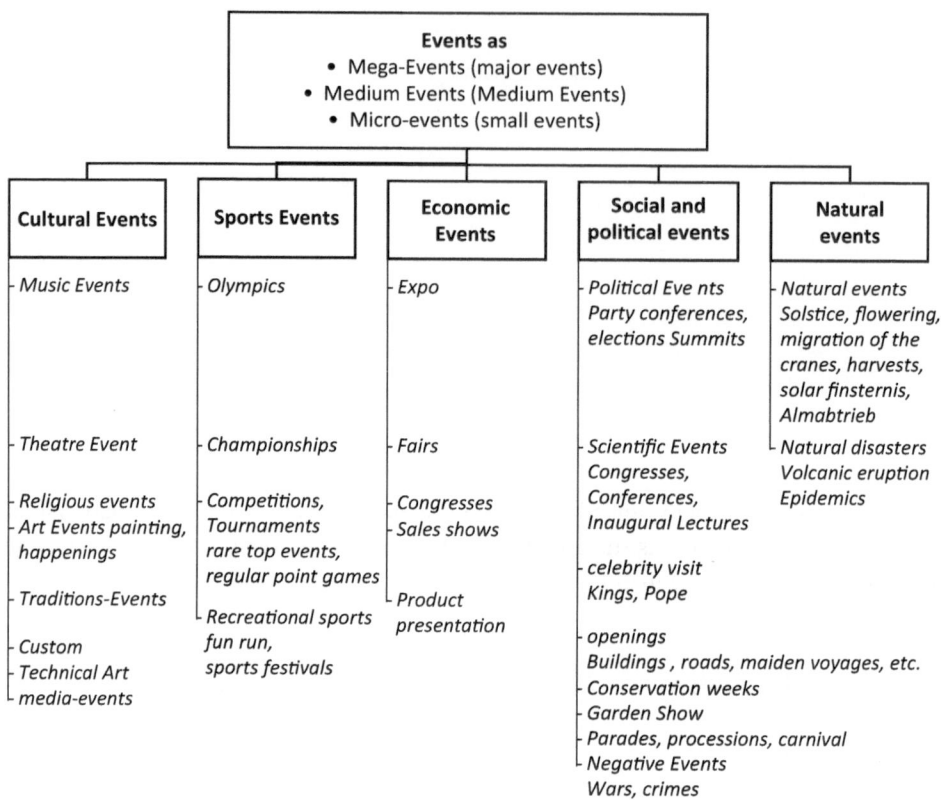

Fig. 5.1 Events at a glance. (Mod. after Freyer 2000, p. 352; Müller 2011, p. 61; own representation and slightly edited by the author)

Many large events such as festivals or open-air music events could not take place in 2020 and 2021 due to Corona. Many theaters and concert halls were closed for months. Many people working in the event industry looked for other jobs. They sometimes even found ones with more regular working hours and higher wages. Experts in the event industry assumed that there might be a shortage of staff after the Corona pandemic, although thousands of jobs were lost in this area (2021a, p. 11). According to a survey by Expo-Event, the Swiss Association of Trade Fair Organizers and Suppliers, almost 4500 jobs were lost in comparison to 2019, which represented a decline of 20%. Many events were postponed in 2020 and 2021—and organizers in Switzerland reported in April 2021 a backlog of orders until 2025 (2021a, p. 11).

5.2 Overtourism

The term *"overtourism"* apparently first appeared on Twitter in 2012 in the form of a hashtag (#overtourism) and in 2016 Rafat Ali from the consulting firm skift.com wrote about *"overtourism"* in an online article (Stettler et al. 2020, p. 70). With *Overtourism* we mean the negative effects of (too) rapid growth in tourism on individual destinations. *Overtourism* or over-tourism can be defined as the amount of tourism and tourist offers that exceed the carrying capacity of a place or city (Visentin and Bertocchi 2019, p. 24). Accordingly, *Overtourism* shows itself on the one hand in a physical overload—too many people in one place at a certain time without control or regulation of the flow of visitors—and on the other hand in the psychological perception by the residents—the feeling of being restricted or restricted by tourism or tourists (Visentin and Bertocchi 2019, p. 30). Often this is accompanied by a local depopulation or at least a decline in the year-round resident population on site. Over-tourism can also lead to the tourist gentrification[1] of individual quarters or places, such as in Palma de Mallorca by tourist restaurants, hotels, souvenir shops, etc., where up to 12 million visitors arrive each year (cf. Blázquez-Salom et al. 2019, p. 39). On the other hand, the tourist gentrification can force residents without basic or home ownership to leave the centers or locations frequented by tourists because they can no longer afford the expensive apartment rents or because there are no more rental apartments (Frenzel 2019, p. 72).

Overtourism—some also speak of *overcrowding*[2]—is increasingly the cause of conflicts in heavily visited destinations between locals and visitors. The locals increasingly perceive the tourists as a nuisance and a burden on everyday life on site. In turn, tourists themselves perceive the large number of co-tourists present as negative or disturbing. Cities with point or broad *Overtourism* are Venice, Barcelona, Dubrovnik, Palma de Mallorca, Amsterdam, Berlin, Lucerne, etc. (Travel Law 2020). However, according to Stettler et al. (2020, p. 71), it is not yet clear where the boundaries are and when *Overtourism* can be spoken of, because there are no universally applicable limit values for this.

According to Peeters et al. (2018, p. 15), *Overtourism* can be defined as follows: "Overtourism describes the situation in which the impact of tourism, at certain times and in certain locations, exceeds physical, ecological, social, economic, psychological, and/or political capacity thresholds". Possible criteria for overtourism include, among others, the tourism density (number of overnight stays per km^2) and the tourism intensity (overnight stays per inhabitant).

---

[1] Gentrification is the economic upgrading of a city district through its renovation or redevelopment. Gentrification usually leads to the displacement of the previously resident population by more affluent population groups—for example by comfortable living lofts.

[2] However, the term *"overtourism"* is more complex and multifaceted than *"overcrowding"* (Peeters et al. 2018, p. 19).

Kuščer and Mihalič (2019, p. 6) listed five central challenges in connection with over-tourism: first, economic consequences for both locals and tourists in the areas of trade, health, and transportation infrastructure; second, socio-cultural effects such as harmful effects on cultural heritage up to damage and vandalism against existing facilities; third, damage to nature such as to forests, waters, fauna, and flora, for example, through pollution and waste; fourth, impairment of stakeholders and local residents as well as unwanted changes in the neighborhood; fifth, negative experiences of tourists themselves through waiting times, excessive queuing, too many tourists, and growing dissatisfaction.

In 2019—that is, before the Corona pandemic—98 destinations in 63 countries were listed on a world map by Responsible Travel that were confronted with overtourism (Frey 2020, p. 6). The European Union even spoke of 105 places in Europe (Kamm-Sager 2019 as well as Frey 2020, p. 6).

A European-wide study of 41 case studies of overtourism found, among other things, that the effects of overtourism can be potentially severe and can lead to both natural and cultural heritage sites at tourism destinations losing their appeal as desirable tourism destinations[3] (Peeters et al. 2018, p. 17). The same study concluded that most tourism destinations are managed primarily from a growth perspective, focusing solely on growth in tourist numbers, without taking into account capacity and other public interests (Peeters et al. 2018, p. 17)[4].

Well-known hotspots of overtourism are, for example, Venice, Mallorca and especially the island capital Palma de Mallorca[5], the Spanish cities of Barcelona, Bilbao, San Sebastian and Madrid, Dubrovnik, Amsterdam, individual districts of Lisbon, Salzburg, Berlin and Hamburg, Hallstatt in the Salzkammergut and Passau. Due to the demand from Asian tourists, Iceland has also become a hotspot for overtourism, as well as certain places in France, some Norwegian fjords and specific attractions such as the Greek island of Santorini, the Italian Lake Garda, Machu Picchu in Peru, the Taj Mahal in India or the Trolltunga rock formation in Norway (Kirstges 2020, pp. 106 ff.).

However, Wheeler (2019, p. XVII) is not wrong to point out that many of the problems associated with overtourism are essentially expressions of "undermanagement". So many local authorities simply do not fulfill their essential control function.

But not everything can be solved at the management level. For example, Rickly (2019, p. 57) has pointed out that the popularity of a destination on social media can lead

---

[3] "The effects of overtourism are potentially severe and both natural and cultural heritage sites are at risk of losing their appeal as desirable tourism destinations due to overtourism" (Peeters 2018, p. 17).

[4] "Most destinations are managed based on a growth-paradigm, mainly valuing growth of visitors' numbers, without considering carrying capacity and other policy goals" (Peeters et al. 2018, p. 17).

[5] For example, in 2017 alone, more than 800 cruise ships called at the Balearic Islands (Marti and Kolly 2019, p. 22).

directly to overtourism, which can promote resentment among the local population and damage the attractiveness of the destination and especially its authenticity. So it is of little use if tourism companies beg tourists to behave ethically and not to post additional photos of fragile ecosystems or vulnerable destinations on Facebook or Instagram (Rickly 2019, p. 57).

Barcelona has long been one of Europe's hotspots for overtourism. The big tourist jump took place in the city with 1.6 million inhabitants in 1992 with the Olympic Games (Martín et al. 2018, pp. 6–7). While 1.73 million tourists visited the city and stayed in hotels in 1990, there were 9 million in 2016. With the other accommodation options included, Barcelona counted 30 million visitors per year before the Corona pandemic. This made Barcelona the 12th most visited city in the world and the third most visited city in Europe after London and Paris (Martín et al. 2018, p. 7). However, only slightly more than 50% of tourists stayed in hotels or hotel apartments. Between 1990 and 2017, the number of overnight stays rose from 18,569 to 67,640, and the number of hotels from 118 to 408 (Martín et al. 2018, p. 7). The tourism intensity was 9807 overnight stays per 1000 inhabitants in 2016, twice as high as the EU average (5209 overnight stays per 1000 inhabitants).

In 2019, Barcelona counted 13 million tourists (Müller 2020, p. 4).

In a survey of residents of Barcelona, 38.6% of those surveyed expressed criticism of tourism and 54.4% had a positive view (Fig. 5.2).

At the same time, 88.6% of those surveyed believed that the attitude towards tourism had changed in the last five years, only 11.4% were of the opposite opinion. And 94% noted an increase in negative attitudes towards tourism (Martín et al. 2018, p. 8).

Martín et al. (2018, p. 13) came to the conclusion in their study that the more critical attitude towards tourism had primarily economic reasons and less sociocultural causes. In addition, the authors believed that booking platforms such as Airbnb were to blame for the growing resentment. Therefore, the growing number of tourists must be controlled and regulated by a corresponding tourism policy (Martín et al. 2018, p. 13).

Fig. 5.2 The attitude of the people of Barcelona towards tourist activities in their city. (From Martín et al. 2018, p. 8; own representation)

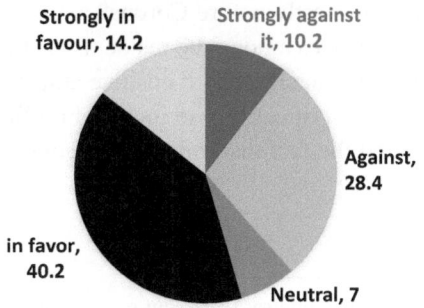

"ARE YOU FOR OR AGAINST TOURIST ACTIVITIES IN THE CITY OF BARCELONA?" (ANSWERS IN % OF RESPONDENTS 2017)

Strongly in favour, 14.2

Strongly against it, 10.2

Against, 28.4

in favor, 40.2

Neutral, 7

Table 5.1 Development of some key figures for the city and tourism in Venice between 2007 and 2017. (From Visentin and Bertocchi 2019, p. 24)

	2007	2017	Change in %
Population	60,755	53,835	−11.4
Arrivals	2,165,656	3,156,000	+46
Overnight stays	5,875,370	7,862,000	+34
Number of hotels	249	274	+10
Hotel rooms	16,015	18,384	+15
Number of restaurants	395	370	−6
Number of stores	2605	2035	−22

Not a few Barcelona residents saw the drop in tourism due to the coronavirus as a great opportunity. For example, Pere Mariné said during the Covid-19 crisis: "We now have the unique opportunity to change something" (quoted after Müller 2020, p. 4). And even Marian Muro, head of the Barcelona Tourism Promotion Agency, said: "We have to make peace with the city" (quoted after Müller 2020, p. 4). But the individual interests are too far apart, the wishes clash too violently.

Even *Venice* was and is today an example of a city shaped by overtourism.

Table 5.1 shows the development of some key figures for the city and tourism between 2007 and 2017.

In 2019, 12 million overnight stays were recorded in Venice (Wysling 2020, p. 4). Officially, 50,000 people still lived in this city this year, but in reality it was probably less—and every year 1000 residents of the lagoon city moved away (Wysling 2020, p. 4). Although Corona had given the city a breather, the big question remains how the culture city, which is burdened by mass tourism, will develop in the future. In principle, there is only one answer: residents instead of tourists. While around 200,000 people live on the mainland, the city of Venice is becoming more and more a tourist shell of hotels, museums, restaurants and souvenir shops—with trampled paths for day-trippers and cruise tourists, for whom even a ship's pier for cruise passengers was built at the Fondamente Nove, in a slightly remote area of the city near the hospital (Wysling 2020, p. 4).

A survey in *Amsterdam* in 2015 showed various hotspots for overtourism (Fig. 5.3).

In *Lucerne* shortly before Corona[6] a representative survey of the Lucerne University of Applied Sciences among the residents of the city showed that although the majority of the population generally had a positive attitude towards tourism, there were a number of points that were critical. These included traffic problems, rising housing prices, the limited space and the fact that only a few people benefit from tourism (Dähler 2020, p. 17).

---

[6] The written questionnaires were sent out in January and February 2020 (see Dähler 2020, p. 17).

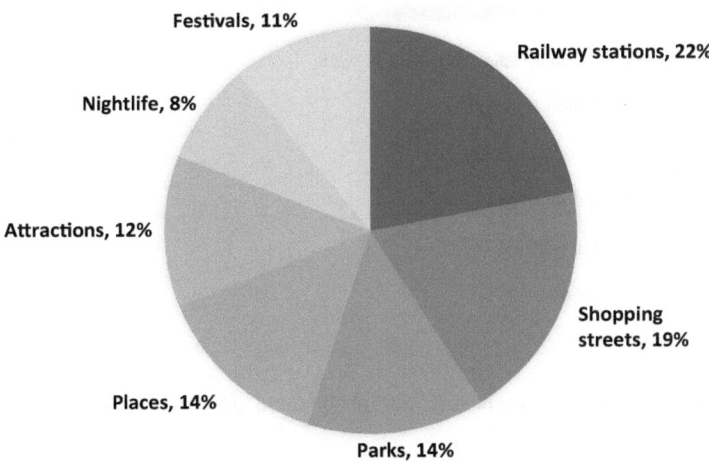

Fig. 5.3 Hotspots for tourist concentrations according to a survey in Amsterdam. (From Gerritsma 2019, p. 138; own representation)

Also criticized were larger and organized tour groups[7], especially from Asia—as well as car tourism. Nevertheless, 54% of those surveyed said that the "acceptable number" of tourists had been exceeded, and in the old town almost 80% of those surveyed were of this opinion (Dähler 2020, p. 17). It was suggested from hotel circles that a fee should be charged for group tourists who drive into the city center by bus and stay there for one or two hours, in order to limit the number of cheap tourists (Zemp in Erni 2018, p. 21). In contrast, tourists who stay in Lucerne for a few days and stay overnight should be addressed in particular. This proposal seemed all the more questionable when a study on tourist value creation in the city of Lucerne showed that of the tourist revenue of 403 million francs in 2017, more than half, namely 224 million francs, was generated in a few jeweler's shops, watch shops and souvenir shops around the Schwanenplatz (Küttel 2018, p. 21). A further 179 million francs fell in retail—especially again in souvenir shops and watch shops—at mountain railways and similar tourist hotspots. Only 3.5% or around 14 million francs went to the mountain railways themselves. A total of 1.4 million group tourists from Asia and the USA had visited the district around the Schwanenplatz. The conclusion of Jürg Stettler, head of the Institute for Tourism at the Lucerne

[7]However, Marcel Perren, director of Luzern Tourismus, said that group tourism also had advantages, such as better control than individual tourism (Perren 2020, p. 17).

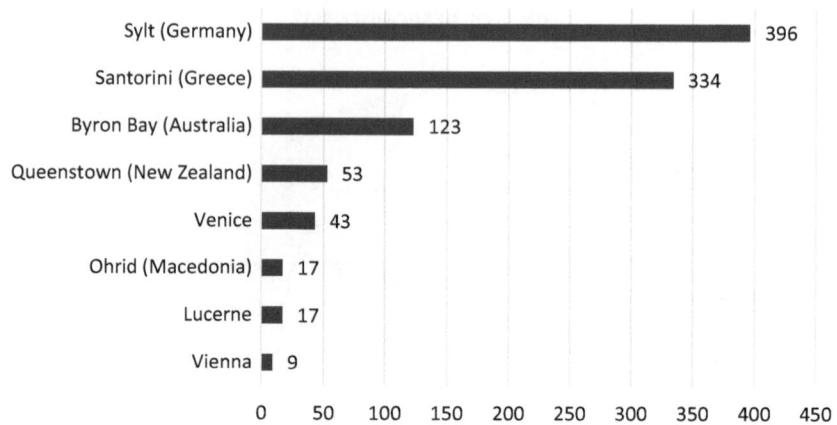

Fig. 5.4   Number of tourist overnight stays per inhabitant in 2018. *Source* Weber (2019, p. 9); own representation

University of Applied Sciences, was: "The majority of the revenue is generated by a few players" (quoted after Küttel 2018, p. 21).

Based on studies in three EU cities, namely Barcelona, Berlin and Venice, Milano (2017 and 2018) has identified several elements that can lead to dissatisfaction due to overtourism:

- Traffic jams or congestion in public places in city centers,
- Privatization of public spaces,
- Increase in land prices,
- Massive increase in the number of cruise ships and high tourist numbers in a short period of time,
- Loss of purchasing power of residents,
- Imbalance between residents and visitors,
- Commercial gentrification[8] and
- Deterioration of the environment, e.g. the waste situation, noise, air and water quality (Peeters et al. 2018, p. 28).

However, the number of overnight stays per inhabitant is only suitable to a limited extent to say something about the extent of overtourism. Depending on whether the population of an entire city or a particularly tourist-frequented district is taken, the numbers differ completely (Frey 2020, p. 7). This is shown, for example, in Fig. 5.4, where Vienna and

---

[8] For gentrification, see footnote 1.

Table 5.2 Visitor to resident ratio in 2018 in Lucerne and Venice. (Mod. after Aschwanden 2018; own representation)

City	Number of visitors in 2018	Residents	Visitor to resident ratio
Luzern	9.4 million	81,000	117:1
Venice	25 million	260,000[9]	96:1

Lucerne score very well compared to Santorini or Sylt. However, if the respective district that is particularly frequented by tourists is taken as the basis of comparison—as in Vienna and Lucerne—the situation looks completely different.

The problem is that the measure of overnight stays per inhabitant (tourism intensity) does not apply to two tourist categories, which are decisive for overtourism: first, day tourists and second, tourists who stay outside the city—like thousands of Chinese tourists in the Lucerne region or also many visitors to Venice. If the number of visitors is related to the number of inhabitants, then in 2018 Lucerne even ranked ahead of Venice (Table 5.2).

It is also interesting how individual places and countries have reacted to the problem of overtourism. For example, the Thai government closed attractions and attractions in national parks during non-peak times to give ecosystems the opportunity to recover (Hess 2019, pp. 119–120). In addition, the Thai government closed 66 locations in 147 national parks either seasonally or permanently in 2018. Among these closed locations were 24 islands—including 5 forever –, 53 waterfalls, 17 caves, 11 paths and 43 other places. Hess (2019, p. 120) rightly criticized that the temporary closure of tourist destinations would not help if they were flooded by tourists after reopening. To avoid this, the following measures are possible, among others: higher entrance fees, stricter controls on the behavior of tourists and possibly fines and maximum quotas for tourists.

Perkumiene and Pranskūniene (2019, p. 6) found that in situations of overtourism, the mobility rights of tourists and the rights of the local population collide. The right to free movement of tourists cannot simply be equated with an automatic right of entry and stay in the destination country. Local residents have a right to life and free self-determination. The interests of local residents who have nothing to do with tourism and the interests of local businessmen who live mainly from tourism can differ greatly under certain circumstances. In addition, many tourist activities are carried out without permission or authorization, such as the rental of apartments to tourists via platforms such as Airbnb, which may take place against the will of the landlords and to the detriment of other tenants. Overtourism is always a question of degree—this is shown, for example, in the number of transactions, the number of tourists and the resulting conflicts. Perkumiene

---

[9] However, significantly fewer people live in Venice itself without the surrounding area (Table 5.1)—so the ratio of visitors to residents is likely to be much higher than in Lucerne.

and Pranskūniene (2019, p. 13) come to the conclusion that the relationship between the rights of tourists and the right to life of the local population is closely related to the concept of freedom of movement—for both, with the rights of tourists being better protected and also enforced than the rights of the local population.

Overtourism is therefore—quite contrary to Imwinkelried's (2019, p. 12) opinion—anything but a figment of the imagination, neither in Switzerland nor elsewhere.

Tourist companies and state institutions have reacted to the problem of overtourism in various ways (Frey 2020, pp. 29 ff.): With information and appeals to control tourism flows, with increased marketing efforts to better distribute frequencies over time, with temporal and spatial restrictions on visitors, with local bans on apartment rentals, with tax incentives or increased fees, with price increases for visits to attractions and cultural sites, and with the denial of visiting rights. But all of these strategies are either unlikely to be successful or socially unjust: price increases hit less affluent tourists, voluntary appeals are inefficient and quickly overwhelmed by photo reports and selfies on social media, and bans are always the worst solution—at least from an economic point of view.

5.3   Urban Living and City Tourism

Stors et al. (2019, pp. 2 ff.) Have pointed out that the separation of urban everyday life and city tourism as two separate spheres is no longer tenable. So urban life itself can become an object of tourist interest, and vice versa, many activities of city dwellers themselves have a "tourist component". In addition, many urban offers are used just as much by city residents as by tourists—such as many means of transport, leisure activities, museums, cultural events, sports events, etc.

Urry and Larsen (2011, p. 4) had already emphasized 10 years ago that travel and tourist stays take place outside of the normal places of residence and work. So they postulated, so to speak, a separation of "tourist objects" and everyday life. Most tourism researchers assumed a clear distinction between "normal" everyday life on the one hand and "exceptional" travel and tourist stays on the other. In contrast, Larsen (2020, p. 24) now takes the position that the distinction between "normal" everyday life and "exceptional" tourism is outdated.

Tourism expert Marco D'Eramo already called for Florence in 2016 for the number of 18% of the houses that were rented out via Airbnb—mostly to tourists. In Venice it was 9%, in Rome 8%. And the annual growth rate was already more than 20% at that time (D'Eramo 2018, p. 25). In many urban centers in Italy—such as in Rome—many older people already moved to the periphery or to the countryside in a cheaper apartment before Corona, while they rented their apartment in the center for a lot of money. For example, the population of Rome shrank since the 1970s from 3 to 2.8 million in 2018. In the center of Rome, 370,000 people lived after the war, that is, 1945—in 2018 there were less than 80,000 (D'Eramo 2018, p. 25).

Like many other European cities, Barcelona, Berlin, Geneva or Bern, Lisbon also faced the problem of rapidly growing private apartment rentals to short-term tourists via platforms such as Airbnb in 2018. Originally propagated as a way to allow the local population to participate in tourism, the problems quickly accumulated: conflicts in the houses where apartments were rented to short-term guests, increasingly scarce housing for resident city dwellers, rising rents and soaring land and housing prices[10]. In the Lisbon district of Alfama, 20% of the apartments were already rented as *"alojamento local"* in 2018, i.e. as short-term accommodation for tourists (Fischer 2018, p. 26). In some places, the authorities reacted with a numerical restriction of short-term overnight accommodation offers, left-wing parties in Portugal wanted a restriction of these offers to 15% of all rental apartments and to 30% of the total housing stock. Geneva restricted the use of an apartment for short-term tourists to a maximum of 60 days, and a special permit must be applied for in addition (Fischer 2018, p. 26). In other places, landlords of apartments rented to short-term guests require a special insurance, in house communities a higher proportion of short-term renters can be charged for the general costs—for example as a result of stronger use of the lift (Fischer 2018, p. 26).

5.4   Covid-19's Impact on Tourism Infrastructure

The Corona pandemic had a not to be underestimated influence on public transport in many places, both temporarily on the frequencies and on the service offer itself.

As a result of the two-month Corona lockdown in spring 2020, the offer in public transport in France was temporarily reduced to 7% of the normal timetable (Belz and Rasch 2020, p. 19). Several billion euros of state financial aid to the French railway company SNCF were made with specific conditions, such as the modernization of infrastructure, the maintenance of small, little-used lines, the development of night services and the expansion of freight transport. In all these areas, the savings had been set in recent years (Belz and Rasch 2020, p. 19). At the beginning of November 2020, the SNCF had to reduce the train service by 75–80% again, after a new lockdown only allowed travel between regions in urgent cases (cf. Belz 2020, p. 24).

The Corona outbreak also had a significant impact on rail transport in Germany. In March and April 2020, only 10% of the previous passengers used the train, but the timetable was maintained by 75%. This led to one of the worst results of the group since the railway reform 25 years ago (Belz and Rasch 2020, p. 19).

In Switzerland, the largest railway company, the Swiss Federal Railways SBB, transported around 800,000 passengers or a third less than in the previous year as a result of the lockdown. During the first lockdown in March/April 2020, the decline was even 80%

---

[10]For example, in Lisbon and Porto, the prices for housing rose by 20% or more in the first quarter of 2018 alone (Fischer 2018, p. 26).

(Stalder 2020, p. 12). Overall, public transport in Switzerland shrank by around 50% in the Corona year. However, most train connections were maintained—with some timetable thinning.

In many places, offers in public transport, which had previously been partly built up and expanded at considerable cost, were abolished or reduced. For example, in January 2021, in the middle of the second Corona wave, the competent authorities decided to discontinue the railway loading of cars by the Rhaetian Railways between Andermatt and Sedrun via the Oberalp Pass in 2023 because the frequencies were too low and the cost coverage was only just 25%. Up to 2021, two to three car trains had been operated per day (Piazza 2021, p. 23).

The para-hotel industry was also affected by the Corona pandemic. From April to June 2020, collective accommodation in Switzerland recorded a drop in overnight stays compared to the previous year of almost 85%, holiday apartments in the same period by just over 53% and campsites by almost 43% (Von Däniken 2020, p. 23). In addition, the proportion of foreign guests in para-hotel accommodation in Switzerland fell from April to June from 25% to 10% compared to the previous year. However, the accommodation providers were optimistic that they would be able to compensate for this decline in summer, which was partly the case.

As a further consequence of the Corona pandemic, prices for holiday apartments rose sharply in many municipalities in the Swiss Alps. As in other countries, city dwellers probably wanted to secure a second home in the country, in which they could take refuge if necessary and have an alternative to foreign travel. In the Corona year 2020, prices rose by almost 4% compared to 2019, the strongest increase since 2012. A holiday apartment of a higher standard even cost just over 7% more per square metre than in the previous year. Fewer foreign trips, more home office and a higher importance of living were the drivers. In easily accessible holiday destinations, price increases were even higher: in Engelberg by 10%, in the Jungfrau region by 8% and in Davos/Klosters even by 12%. In contrast, price increases in cheaper holiday resorts were only just under 2%—for example in Hasliberg, Adelboden and Wildhaus. In some places, such as Leysin, Crans-Montana and Flumserberg, prices even fell slightly (Jordan 2021b, p. 10).

References

Aschwanden, Erich 2018: Die Angst vor dem „Overtourism". Luzern, Barcelona, Venedig – alle leiden unter demselben Phänomen. In: Neue Zürcher Zeitung vom 2.8.2018. 11.

Belz, Nina 2020: Die Billigbahn Ouigo bringt Franzosen in den Zug. In: Neue Zürcher Zeitung vom 26.11.2020. 24.

Belz, Nina / Rasch, Michael 2020: SNCF und Deutsche Bahn brauchen Hilfe. Die Corona-Krise lässt die Auslastung der Züge sinken – und verschärft die Probleme der Staatsbahnen in Frankreich und Deutschland. In: Neue Zürcher Zeitung vom 24.7.2020. 19.

Blázquez-Salom, Maciá / Blanco-Romero, Asunción / Carbonell, Jaime Gual / Murray, Ivan 2019: Tourist Gentrification of Retail Shops in Palma (Majorca). In: Milano, Claudio / Cheer, Joseph

M. / Novelli, Marina (Hrsg.): Overtourism. Excesses, Discontents and Measures in Travel and Tourism. Oxfordshire/Boston: CABI. 39 ff.

Dähler, Stefan 2020: Bevölkerung will Tourismus regulieren. In: Luzerner Zeitung vom 10.6.2020. 17.

D'Eramo, Marco 2018: „Der Reisende ist nur ein Tourist, der abstreitet, einer zu sein". Gespräch mit Marco D'Eramo von Daniel Weber. In: NZZ Folio Nr. 10 (2018). 24 ff.

Erni, Fritz 2018: Mit Gebühren gegen Car-Touristen. Interview mit Fritz Erni, Direktor des Hotels Montana in Luzern, von Raphael Zemp. In: Luzerner Zeitung vom 11.8.2018. 21.

Fischer, Thomas 2018: Portugal drosselt den touristischen Wildwuchs. In: Neue Zürcher Zeitung vom 14.8.2021. 26.

Frenzel, Fabian 2019: Tourist Valorisation and Urban Development. In: Frisch, Thomas / Sommer, Christoph / Stoltenberg, Luise / Stors, Natalie (Hrsg.): Tourism and Everyday Life in the Contemporary City. London/New York: Routledge. 68 ff.

Frey, Bruno S. 2020: Venedig ist überall. Vom Übertourismus zum Neuen Original. Wiesbaden: Springer.

Freyer, W. 2000: Ganzheitlicher Tourismus. Beiträge aus 20 Jahren Tourismusforschung. Dresden: Verlag TU.

Gerritsma, Roos 2019: Overcrowded Amsterdam: Striving for a Balance Between Trade, Tolerance and Tourism. In: Milano, Claudio / Cheer, Joseph M. / Novelli, Marina (Hrsg.): Overtourism. Excesses, Discontents and Measures in Travel and Tourism. Oxfordshire/Boston: CABI. 125 ff.

Hess, Janto S. 2019: Thailand: Too Popular for its Own Good. In: Dodds, Rachel / Butler, Richard W. (Hrsg.): Overtourism. Issues, Realities and Solutions. De Gruyter Studies in Tourism. Volume 1. Berlin/Boston: De Gruyter. 111 ff.

Imwinkelried, Daniel 2019: Ein Schweizer Hirngespinst namens Overtourism. In: Neue Zürcher Zeitung vom 4.12.2019. 12.

Jordan, Gabriela 2021a: Der Eventbranche läuft das Personal davon. In: Luzerner Zeitung vom 7.4.2021. 11.

Jordan, Gabriela 2021b: Die Nachfrage nach Ferienwohnungen steigt und steigt. In: Luzerner Zeitung vom 25.3.2021. 10.

Kamm-Sager, Christa 2019: (Zu) viele Touristen: Diese Orte sind überfüllt, haben kapituliert oder sind ganz geschlossen für die Massen. In: St. Galler Tagblatt vom 8.10.2019. https://www.tagblatt.ch/leben/zu-viele-touristen-diese-orte-sind-ueberfuellt-haben-kapituliert-oder-sind-ganz-geschlossen-fuer-die-massen-ld.1158124 (Zugriff 24.4.2021).

Kirstges, Torsten H. 2020: Tourismus in der Kritik. Klimaschädigender Overtourism statt sauberer Industrie? München: UVK Verlag.

Kuščer, Kir / Mihalič, Tanja 2019: Residents' Attitudes towards Overtourism from the Perspective of Tourism Impacts and Cooperation – The Case of Ljubljana. In: Sustainaility 11/6 (2019) 1823. www.mdpi.com/journal/sustainability (Zugriff 21.2.2021).

Küttel, Kilian 2018: Die Bergbahnen profitieren nur wenig. In: Luzerner Zeitung vom 26.6.2018. 21.

Larsen, Jonas 2020: Ordinary Tourism and Extraordinary Everyday Life. Rethinking Tourism and Cities. In: Frisch, Thomas / Sommer, Christoph / Stoltenberg, Luise / Stors, Natalie (Hrsg.): Tourism and Everyday Life in the Contemporary City. London/New York: Routledge. 24 ff.

Marti, Gian Andrea / Kolly, Marie-José 2019: Dichtestress wegen Riesenschiffen. In: Neue Zürcher Zeitung vom 5.6.2019.

Martín Martín, José María / Guaita Martínez, Jose Manuel / Salinas Fernández, José Antonio 2018: An Analysis of the Factors behind the Citizen's Attitude of Rejection towards Tourism in a Context of Overtourism and Economic Dependence on This Activity. In: Sustainability 10 (2018) 2851. www.mdpi.com/journal/sustainability (Zugriff 11.1.2021).

Milano, C. 2017: Over-tourism and Tourism-phobia: Global trends and local contexts. Barcelona.

Milano, C. 2018: Overtourism, malestar social y turismofobia. Un debate controvertido. In: Pasos. Revista de Turismo y Patrimonio Cultural, 6/3 (2018). 551 ff.

Müller, Hansruedi 2011: Tourismuspolitik. Wege zu einer nachhaltigen Entwicklung. Glarus/Chur: Rüegger.

Müller, Ute 2020: Barcelona ist ein Schatz – lässt er sich verstecken? In: Neue Zürcher Zeitung vom 1.7.2020. 4.

Peeters, P., Gössling, S., Klijs, J., Milano, C., Novelli, M., Dijkmans, C., Eijgelaar, E., Hartman, S., Heslinga, J., Isaac, R., Mitas, O., Moretti, S., Nawijn, J., Papp, B. and Postma, A. 2018: Overtourism. Impact and Possible Policy Responses. Research for TRAN Committee. Brussels: European Parliament. Policy Department for Structural and Cohesion Policies.

Perkumiene, Dalia / Pranskūniene, Rasa 2019: Overtourism: Between the Right to Travel and Residents' Rights. In: Sustainability 11 (2019) 2138. www.mdpi.com/journal/sustainability (Zugriff 12.1.2021).

Perren, Marcel 2020: Nachgefragt: „Der Gruppetourismus hat auch Vorteile". In: Luzerner Zeitung vom 10.6.2020. 17.

Piazza Matthias 2021: Endstation für den Autozug. In: Luzerner Zeitung vom 14.1.2021. 23.

Reiserecht 2020: Information, Checklisten und Muster zum Reiserecht in der Schweiz. https://www.reiserecht.ch/tourismus/overtourism (Zugriff 13.1.2021).

Rickly, Jillian M. 2019: Overtourism and Authenticity. In: Dodds, Rachel / Butler, Richard W. (Hrsg.): Overtourism. Issues, Realities and Solutions. De Gruyter Studies in Tourism. Volume 1. Berlin/Boston: De Gruyter. 46 ff.

Stalder, Helmut 2020: Corona-Krise pulverisiert SBB-Ergebnis. In: Neue Zürcher Zeitung vom 11.9.2020. 12.

Stettler, Jürg / Eggli, Florian / Weber, Fabian / Huck, Lukas 2020: Overtourism in der Schweiz. Herausforderungen und Lösungsmöglichkeiten. Das Fallbeispiel Luzern. In: Bieger, Thomas / Beritelli, Pietro / Laesser, Christian (Hrsg.): Innovative Konzepte im alpinen Tourismus. Schweizer Jahrbuch für Tourismus 2019/2020. Berlin: Erich Schmidt Verlag. 69 ff.

Stettler Jürg / Wallebohr, Anna / Herzer, Christine / Hoff, Oliver 2016: Bedeutung und Bewertung von Events zur Beurteilung ihrer Förderwürdigkeit. Analyse von vier Sportgross- und einer Megaspott-Veranstaltung in der Schweiz. In: Zeitschrift für Tourismuswissenschaft. Volume 8/ Issue 2 (2016). Themenheft Ökonomische Wirkungen des Tourismus und ihre Einflussfaktoren. Berlin/Boston: De Gruyter. 253 ff.

Stors, Natalie / Stoltenberg, Luise / Sommer, Christoph / Frisch, Thomas 2019: Tourism and Everyday Life in the Contemporary City. An Introduction. In: Frisch, Thomas / Sommer, Christoph / Stoltenberg, Luise / Stors, Natalie (Hrsg.): Tourism and Everyday Life in the Contemporary City. London/New York: Routledge. 1 ff.

Urry, John / Larsen, Jonas 2011: The Tourist Gaze 3.0. Thousand Oaks: Sage.

Visentin, Francesco / Bertocchi, Dario 2019: Venice: An Analysis of Tourism Excesses in an Overtourism Icon. In: Milano, Claudio / Cheer, Joseph M. / Novelli, Marina (Hrsg.): Overtourism. Excesses, Discontents and Measures in Travel and Tourism. Oxfordshire/Boston: CABI. 18 ff.

Von Däniken, Alexander 2021: Gruppen und Schulen fehlen in den Jugis. In: Luzerner Zeitung vom 12.9.2020. 23.

Weber, Fabian 2019: „Widersprüchlichkeiten sind üblich". Interview mit Fabian Weber von Rainer Rickenbach. In: Luzerner Zeitung vom 2.5.2019. 9.

Wheeler, Tony 2019: Foreword. In: Milano, Claudio / Cheer, Joseph M. / Novelli, Marina (Hrsg.): Overtourism. Excesses, Discontents and Measures in Travel and Tourism. Oxfordshire/Boston: CABI. XV ff.

Wysling, Andres 2020: Wie Venedig den Overtourism überwinden könnte. In: Neue Zürcher Zeitung vom 17.9.2020. 4.

Zenhäusern, Robert / Kadelbach, Thomas 2016: 12 Thesen zur Zukunft des Tourismus in den Berggebieten. Bern: Schweizer Tourismusverband / Schweizerische Arbeitsgemeinschaft für die Berggebiete. Juli 2018.

Everyone Wants to Travel—(Almost) No One Wants the Tourists

Kirig (2019, p. 24) has put forward the thesis that today's people are not only looking for short-term recreation or a single moment of happiness on a journey, but "a transformative experience through the exchange with the environment that expands the personal life story". However, such an assessment seems rather detached in a society that is characterized by mobility as never before. Are the motivations for travel not much more prosaic: variety, curiosity about something else or simply escape from everyday life seem to be much more common motivations. Of course it is true that many tourists are looking for a certain way of life on their journey, but to conclude from this that there is a deep contribution to identity formation, self-development and resilience (Kirig 2019, p. 24) is probably somewhat exaggerated. Just as well one can understand every individual encounter or experience as identity-forming—or every event. But you don't have to travel for that—and certainly not as a tourist or tourist.

It also seems a bit naive to see tourism as a means of peace, democracy and intercultural understanding, as Edgell and Swanson (2013, p. 129) did (Fig. 6.1).

Even if the influence of tourism on cultural understanding and democracy is relatively minimized (*dotted arrows*)—the fact is that a causal influence of tourism on democracy and peace has hardly ever been proven. At most, an indirect influence via the generation of income and greater prosperity on the willingness for peace can be assumed.

At the same time, the following question arises: If one—as Muntschik (2019, p. 33) sees mobility as the future normal state and as emancipation from tourism—should one not say goodbye to tourism altogether? If the society of the future is an exclusively mobile society, then tourism as such has become obsolete because then everyone is mobile and can go anywhere at any time. But precisely the recent experience of the Corona pandemic has shown that very little—such as a not even too dangerous virus—is needed, and the whole mobility dissolves into smoke, starting with border closures, over quarantine regulations to lockdowns and the isolation of entire cities or areas.

C. J. Jäggi, *Tourism Before, During and After Corona*, https://doi.org/10.1007/978-3-658-39182-9_6

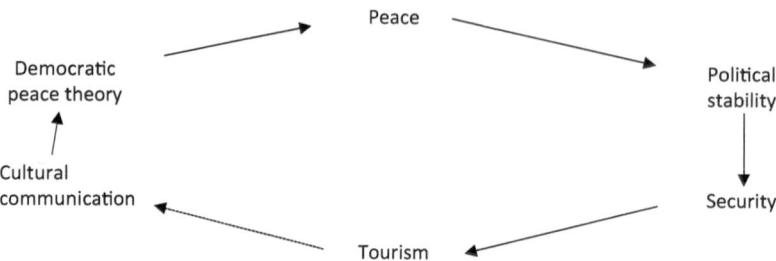

Fig. 6.1 Peace—intercultural understanding security cycle. (Mod. after Edgell and Swanson 2013, p. 12; own representation)

6.1   Tourism, Security and Political Stability

As Scheyvens (2011, p. 11) rightly pointed out, the close connection between political stability, security and tourism is now widely recognized. Political upheavals, terrorism and criminal violence can deal a death blow to tourism destinations overnight or prevent tourism from taking off at all. Studies have shown, for example, that increasing political instability can lead to a decline in tourism of 20–25% in the long term (2011, p. 11), and individual events such as terrorist attacks can deal a death blow to tourism on a temporary basis.

This means that the security theme is decisive for tourism and will probably become even more important in the future. This is all the more true as international tourism partly moves in a more or less lawless area or in legal grey areas which are only insufficiently covered by national jurisdiction.

6.2   What Tourism Policy?

According to Bandi Tanner et al. (2018, p. 164), tourism policy can be understood as the targeted promotion and shaping of tourism through influencing tourist-relevant conditions, in the sense of overarching state goals and in the case of obvious market failure. This is about strategic positioning of tourism, increasing strategic capability, improved market appearance, reducing potential for conflict and promoting attractiveness of the location.

In a much-cited article, Fayos-Solá (1996) listed three phases or generations of tourism policy: In a first phase, efforts were focused on increasing the volume of tourism in the sense of "more of the same", in a second phase attempts were made to expand tourism as an industry through contributions, subsidies and regulations in the sense of spatial planning and environmental requirements according to the motto "more of the most", and in a third phase the competitiveness should be promoted by higher quality and efficiency, in the sense of "more of the best" (Henriksen and Halkier 2012, p. 7).

But it is not clear whether the third step has taken place everywhere—many tourism providers and destinations are still oriented towards the goal of unlimited quantitative growth in the sense of "the more the better".

A tourism policy in another sense was pursued by Russian President Putin towards Turkey: After around 7 million Russians spent their holidays in Turkey in 2019, Russia stopped all flights to Turkey in April 2021. Officially, the epidemiological situation of Covid-19 in Turkey was given as the reason, but observers were of the opinion that the actual reason was the increased arms cooperation of Turkey with Ukraine—especially since Russia had already used tourism in previous years to exert pressure on the Turkish government (Pabst 2021, p. 3).

6.3   Crisis Management in Tourism

Until 2019, crisis communication in tourism was mainly discussed from the perspective of terrorism (Boven 2019, pp. 19 ff.), against the background of accidents and disasters such as the Costa Concordia (Hahn 2018, p. 36) or from the perspective of natural disasters such as the tsunami in Southeast Asia in 2004, avalanches or floods (Hahn 2018, p. 43).

As typical crisis courses, Hahn (2018, pp. 42 ff.) named three types of crises: 1.) A crisis due to a single and surprising initial event, 2.) a "creeping" crisis that lasts for a long time and only reaches its peak after a while, and 3.) a "crisis wave" that can rise and fall over a longer period of time. The first type of an "eruptive" crisis, for example as a result of a catastrophe or a scandal as a result of a real or media initial event, immediately attracts high media attention and can disable the everyday business of an organization in a very short time. The second type of crisis develops slowly, is often noticed late and, in retrospect, should have been seen coming. Media attention is initially low, increases and often lasts for a long time. Such crises can be caused by logistics or service deficits as well as by infrastructure problems (traffic), and public interest depends on which multiplier groups get involved in the discourse. The third type of crisis, a wave-like crisis, fluctuates up and down, media attention is in the middle range, but periodically focuses attention on the relevant organization, institution or the relevant issue. A crisis of the first type can also turn into a crisis of the third type, such as the Covid-19 pandemic.

Berg (2014, p. 76) has distinguished between potential crises, latent crises, acute but controlled crises, and acute but uncontrolled crises. Potential crises only exist in thoughts and are not yet real; latent crises have indeed broken out, but their causes have not yet been recognized according to Berg (2014, p. 76), acute but controlled crises are also real, but a positive outcome is expected, and acute, uncontrolled crises can no longer be managed, for example, a liquidation by order of the court.

Berg (2014, p. 77) has also distinguished between external and internal causes of crises on the one hand and exogenous and endogenous causes on the other (Table 6.1).

Table 6.1 External, internal, exogenous and endogenous causes of crises. (From Berg 2014, p. 77)

External causes	Cyclical downturns, changes in consumer behavior, natural disasters, terrorism, political instability, epidemics / pandemics, wars and unrest
Internal causes	Management errors, quality defects, security defects
Exogenous causes	Crisis causes in the target areas: e.g. geophysical, socio-cultural, political, religious and health factors (diseases); Crisis causes on the journey: e.g. hijackings, attacks on travel buses, trains
Endogenous causes	Crisis cause human: e.g. management error; Qualification deficiencies of personnel: e.g. no or only insufficient safety knowledge, negligence, disregard of regulations, strikes, sabotage; Crisis cause technical: e.g. missing or inadequate safety precautions, cost-related reduction of safety, technical failure due to material or construction defects, wear

Table 6.2 Material and immaterial effects of crises on companies or industries. (From Berg 2014, p. 77)

Material impacts	Immaterial impacts
• Cancellations and rebookings • Decrease in utilization • Decreasing sales • Additional costs for removal, compensation • Capital loss	• Image damage to the company and to individuals • Loss of trust • Customer attrition • Loss of motivation in staff

However, the distinction between external and exogenous and internal and endogenous[1] is not really convincing because it is not clearly delineated.

In addition, the effects of crises on companies, including in the tourism sector, can have both material and immaterial effects (Table 6.2).

6.4   Tourism as a Search for Meaning—A New Market Niche?

Innerhofer and Pechlaner (2016, p. 12) have argued that pilgrimage and pilgrimage travel are among the oldest forms of tourism, while the search for spirituality and the connection of spirituality and tourism have only gained importance in recent decades. Against

---

[1] Steinbach (2003, p. 45) has distinguished between "endogenous" reproduction of tourist perspectives within their original cultural space in the form of "local" equipment of tourist enclaves and "exogenous" reproduction, i.e. "export" of tourist perspectives into foreign cultural spaces, for example as "implementation of replicas in enclaves at distant locations". An example of the latter would be the tourist reproduction of Venice in fin-de-siècle Vienna around 1880 on the grounds of the Prater or on the grounds of the hotel-casino "The Venetian" in Las Vegas.

the background of growing individualization, Innerhofer and Pechlaner (2016, p. 13) identify a stronger distance from traditional religious forms and a marginalization of religion. At the same time, however, there is an increased search for alternative lifestyles and spiritual experiences. Tourism tries to pick up on the spiritual yearning through various forms of travel and to meet the desire for relaxation, tension and the supernatural through corresponding offers and products. In the opinion of Innerhofer and Pechlaner (2016, p. 15), it is "less about experience and fun orientation and more about experience and searching for meaning, less about staging and more about authenticity". This is certainly worth considering.

However, from a theological and spiritual point of view, there are two things to consider: True spirituality means a holistic, radical commitment to inner or transcendental values. However, this can only happen to a limited extent through events or tourist offers, which, alongside other interests, simply also cover religious or spiritual consumption needs. The strong need for vacation, travel or holidays is, among other things, also the result of our unspiritual way of life, our materialistic lifestyle or our quest for always new exciting experiences[2].

6.5 Volunteer Tourism

The roots of so-called volunteer tourism ("*volunteer tourism*") go back to the nineteenth century, when missionaries, doctors and teachers traveled to help others (Benson 2011, p. 1).

About 10 years ago, the expenditures generated by volunteer tourism were estimated at an amount between US$1.66 and US$2.6 billion, with some 1.6 million "volunteer tourists" (Benson 2011, p. 1). This type of tourism is often associated with relief organizations, some of which operate in the non-profit sector, some in the form of social entrepreneurship, and some purely commercial.

Volunteer tourism combines travel with assistance to the poor and disadvantaged, with commitment to environmental or social issues (Alexander and Bakir 2011, p. 9).

There is no doubt that the desire to contribute to the improvement of the living and environmental conditions at the destination is positive about this type of tourism. Alexander and Bakir (2011, p. 15) have listed various forms of "engagement" (Table 6.3).

---

[2] To put it bluntly: It is not about a spiritual supermarket where you can buy 300 g of yoga, 2 kg of meditation, 50 g of fire walking, 400 g of church visits, 100 g of shamanic healing rituals, 300 g of temple visits and 100 g of *sacred dance*. Because spirituality cannot be consumed or enjoyed—rather, spirituality happens and happens as a deep experience and development process. Therefore, it is about more than just a "form of travel for recreation and relaxation, which is fed by the yearning for post-material values, for healing expectations, ideas of wholeness and a yearning for the supernatural" (Innerhofer and Pechlaner 2016, p. 17)—best of all with the booking of a "spiritual coach" (Innerhofer and Pechlaner 2016, p. 22). Here, tourist offers clearly reach their limits.

Table 6.3 Forms of engagement in volunteer tourism. (Mod. nach Alexander und Bakir 2011, p. 15; Translation from English and slightly modified by author)

Core category Engagement	
Concepts	Property, activity
Participation	Participate in an activity
Action	Do something
Integration	Mixing or connecting with other people or ethnic groups
Penetration	Seeing clearly and deeply
Interaction	Influencing and exchanging with each other
Involvement	Being connected or joined with someone
Dip	Be caught and absorbed

But are the damages caused by tourists really compensated by such an attitude and corresponding actions? Isn't youth idealism simply being commercialized here? You have to look closely at that.

In any case, organizations such as voluntouring.org (2021), which boasts the title "volunteering and meaningful travel around the world", or volunteerworld.com appear more like lightly social or green-tinged travel marketers than serious providers of social or environmental engagement[3].

6.6   Global Code of Ethics for Tourism

In response to growing criticism of individual phenomena and impacts of global tourism, the UNWTO General Assembly adopted a Global Code of Ethics for Tourism in 1999 (Global Code of Ethics for Tourism 1999; Coghlan 2019, pp. 145 ff.). Among other things, it was stipulated that tourism should lead to mutual understanding and respect (Article 1), that tourism should be a vehicle for individual and collective fulfilment (Article 2), that tourism should be understood as a factor for sustainable development (Arti-

---

[3] For example, volunteerworld.com (2021) advertises its offer and its 66 programs with the following sentences: "Do you dream of backpacking through some of the most remote places abroad but somehow feel as if it's not enough? Do you want to do more than that? Do you want to connect to the communities and help? If you are interested in volunteering abroad and you have already done some research on the topic, you might have come across the term 'Voluntourism'. It simply describes 'tourism in which travelers do voluntary work to help communities or the environment in the places they are visiting', but in the past years, there has been a lot of discussion about the pros and cons of voluntourism abroad."

cle 3), that tourism should preserve the cultural heritage of humanity (Article 4), that tourism should be a benefit for host countries and communities (Article 5), that all stakeholders in tourism development are obliged (Article 6), that there is a right to tourism (Article 7), that tourists have a right to free movement in their own country and to visit other countries (Article 8), that the rights of all those working in tourism are to be guaranteed (Article 9) and that all these rights are to be enshrined in a global code of ethics for tourism (Article 10).

According to Baleva (2019, p. 210), the first article emphasises the goal and importance of tourism for mutual understanding, while articles 2–6 and 9 contain ethical principles for the private sector. Article 3 emphasises the role of tourism for development and Article 5 speaks of the benefits for host countries. Articles 6–8 emphasise the rights of tourists and stakeholders, that is, all those affected by tourist offers.

6.7   Travel in the Corona Era

As early as January 2021, that is, during the Corona period, cruise companies advertised with the predicate "coronafree". This is how an advertisement of the company MSC sounded: "Worry-free on a cruise", and the company presented its ships as a "safe bubble". In 2020, as with the rest of tourism, the cruise business also collapsed (Jordan 2021, p. 15). Not a few cruise ships had to shorten their journey in spring 2020, could no longer dock at destinations and in some cases a quarantine was imposed on cruise ships. From the end of 2020, interest in cruises increased continuously—the travel agencies explicitly marketed "safe cruise tours" without intermediate stops and landings and with extended protection concepts. But the practice looks different even with these so-called "blue tours": The passengers can indeed go ashore and "discover nature and culture" (Jordan 2021, p. 15)—provided that the destination country or the port authorities do not deny them the landing. So it's all hot air and a marketing gag? After all, even passengers who stay on the ship throughout the journey need an airplane, train or bus to get there—all of which are also opportunities to catch the virus.

Two things were noticeable in the context of winter tourism in Austria and Switzerland: On the one hand, the open ski areas did not prove to be hotspots for Covid-19 (Benz 2021, p. 21)[4]. After Christmas 2020 and New Year 2021, the number of Corona cases decreased in a number of countries, including Switzerland. Thanks to mass testing and protection concepts, hotspots could be detected early, for example in two hotels in St. Moritz, as well as in Wengen and Arosa. But on the other hand, tourists and catering staff—many of them not locals—proved to be transmitters of the virus. Paradoxically,

---

[4] The infamous example of Bad Ischgl in spring 2020 is not necessarily a counter-example, because the infections there did not occur on the ski slopes, but during the après-ski—and in the later course of the Covid-19 pandemic, the bars and restaurants were closed.

the restaurants were closed for locals, but the hotels were allowed to serve their guests. Although in the winter season 2020/2021 only around 10–15% of hotel guests in Switzerland came from abroad, the rest from their own country. A hotelier from Lenzerheide said about this: "Frustration and aggression have built up in the population in recent months. A stay in the mountains can be a valve" (quoted according to Benz 2021, p. 21). However, one would have to add: also a new possibility of infection.

But summer tourism was also severely affected by the Corona crisis in 2020. For example, in the central Swiss city of Lucerne, tourists from Asia generated 32% of overnight stays in the pre-Corona peak year 2019, and tourists from the USA 20%. In comparison, in the same year, Asian tourists generated 14% of overnight stays and US visitors 6% of overnight stays nationwide (Aschwanden 2021, p. 9). In 2020, overnight stays fell by 65% compared to the previous year, and for tourists from Asia by 90.2%. When one considers that tourism in the city of Lucerne generated a value added of 760 million Swiss francs and around 7500 jobs in 2019—or, in other words, 7.4% of the economy or 12.2% of all jobs (Aschwanden 2021, p. 9)—it is easy to see how severe the economic consequences of the tourist collapse were in this city alone.

Covid-19 also led to significant travel restrictions in the Pacific region. For example, travel between Australia, New Zealand and Fiji had been suspended, and Fiji was campaigning for a so-called "Bula Bubble" in spring 2021, the creation of a common zone between Australia, New Zealand and Fiji, in which travel would be made easier despite Covid-19. Similar travel facilitation already existed between Australia and New Zealand, which was withdrawn in the event of local outbreaks. Entry to Fiji was generally possible in February 2021, but only with a screening of arriving passengers for Corona symptoms and a state-controlled quarantine in a quarantine facility (Pazifik aktuell March 2021, p. 14).

Japan took a different approach: This country, which was visited by almost 32 million tourists in 2019, subsidized travel within the country, hotel stays, the purchase of theater and concert tickets, and the purchase of souvenirs for Japanese people. So the state took over a third of the accommodation costs—and up to 20,000 yen, or around 150 € could be saved per person and night. By mid-November, 52.6 million travelers had already benefited from this offer (Putz 2020, p. 3). However, the campaign was suspended from December 28, 2020 to January 11, 2021 due to high infection rates.

References

Alexander, Zoë / Bakir, Ali 2011: Understanding Voluntourism. A Glaserian Grounded Theory Study. In: Benson, Angela M. (Hrsg.): Volunteer Tourism. Theoretical Frameworks and Practical Applications. London/New York: Routledge. 9 ff.

Aschwanden, Erich 2021: Luzern sucht Alternativen zu chinesischen Reisegruppen. In: Neue Zürcher Zeitung vom 4.5.2021. 9.

Baleva, Mary Kristerie A. 2019: Regaining Paradise Lost: Indigenous Land Rights and Tourism. Leiden/Boston: Brill Nijhoff.

Bandi Tanner, Monika / Pfammatter, Adrian / Weber, Romina / Lehmann Friedli Therese 2018: Förderung der Strategiefähigkeit touristischer Unternehmen durch nationale Tourismuspolitiken. In: Zeitschrift für Tourismuswissenschaft. Volume 10/Issue 2 (2018). Themenheft Internationalisierung des Tourismus – Tourismus im Wandel. Oldenbourg: De Gruyter. 161 ff.

Benson, Angela M. 2011: Volunteer Tourism. Theory and Practice. In: Benson, Angela M. (Hrsg.): Volunteer Tourism. Theoretical Frameworks and Practical Applications. London/New York: Routledge. 1 ff.

Benz, Matthias 2021: Die Schweizer Skigebiete dürften offen bleiben. In: Neue Zürcher Zeitung vom 1.2.2021. 21.

Berg, Waldemar 2014: Modul A: Einführung Tourismus. Überblick und Management. Kapitel 3: Ausgewählte Managementformen im Tourismus. In: Schulz, Axel / Berg, Waldemar / Gardini, Marco A. / Kirstges, Torsten / Eisenstein, Bernd: Grundlagen des Tourismus. Lehrbuch in 5 Modulen. 2., überarbeitete Auflage. München: Oldenbourg Verlag. 63 ff.

Boven, Christine 2018: Tourismus und Terrorismus und die Rolle von Risikowahrnehmung: Forschungsansätze. In: Hahn, Silke / Neuss, Zeljka (Hrsg.): Krisenkommunikation in Tourismusorganisationen. Grundlagen, Praxis, Perspektiven. Wiesbaden: Springer VS. 20 ff.

Coghlan, Alexandra 2019: An Introduction to Sustainable Tourism. Oxford: Goodfellow Publishers.

Edgell, David L. / Swanson, Jason R. 2013: Tourism Policy and Planning. Yesterday, Today, and Tomorrow. London/New York: Routledge.

Fayos-Solá, E. 1996: Tourism Policy: A Midsummer Night's Dream? In: Tourism Management. 17/6 (1996). 405 ff.

Global Code of Ethics for Tourism 1999: The World Tourism Organization UNWTO. https://webunwto.s3.eu-west-1.amazonaws.com/imported_images/37802/gcetbrochureglobalcodeen.pdf (Zugriff 20.1.2021).

Hahn, Silke 2018 Krisenmanagement und Krisenkommunikation: Phasen, Zielgruppen, Wirkhebel und das Prinzip Hoffnung. In: Hahn, Silke / Neuss, Zeljka (Hrsg.): Krisenkommunikation in Tourismusorganisationen. Grundlagen, Praxis, Perspektiven. Wiesbaden: Springer VS. 35 ff.

Henriksen, Pennie F. / Halkier, Henrik 2012: From Local Promotion Towards Regional Tourism Policies: Knowledge Processes and Actor Networks in North Jutland, Denmark. In: Kumral, Neşe / Önder, A. Özlem (Hrsg.): Tourism, Regional development and Public Policy. London/New York: Routledge. 5 ff.

Jordan, Gabriela 2021: Luxus-Schiffe werben mit „coronafrei". In: Luzerner Zeitung vom 23.1.2021. 15.

Kirig, Anja 2019: Resonanz als transformative Erfahrung ist das Grundbedürfnis des Menschen in einer wir-kulturellen Gesellschaft. In: Zukunftsinstitut. Trendstudie: Der neue Resonanz-Tourismus. Herzlich willkommen. Frankfurt/Main: Zukunftsinstitut. 24 ff.

Muntschik, Verena 2019: Exkurs: Resonanz. Ein neues gesellschaftliches Bedürfnis nach Beziehungserfahrungen. In: Zukunftsinstitut. Trendstudie: Der neue Resonanz-Tourismus. Herzlich willkommen. Frankfurt/Main: Zukunftsinstitut. 26 ff.

Pabst, Volker 2021: Der Tourismus hat oberste Priorität in der Türkei. In: Neue Zürcher Zeitung vom 12.5.2021. 3.

Pazifik aktuell März 2021: Covid-19: Fidschi plädiert weiter für „Bula Bubble" mit Australien und Neuseeland. Nr. 125 (2021). 14.

Pechlaner, Harald / Innerhofer, Elisa 2016: Spiritualität & Tourismus – Schnittstellen und Perspektiven. In: Pechlaner, Harald / Innerhofer, Elisa (Hrsg.): Sinnsuche im Urlaub. Chancen und Perspektiven für den Tourismus. Bozen: Athesiaverlag. 11 ff.

Putz, Ulrike 2020: Balkonien statt Binnentourismus in Japan. In: Neue Zürcher Zeitung vom 21.12.2020. 3.

Scheyvens, Regina 2011: Tourism and Poverty. New York: Routledge.

Steinbach, Josef 2003: Tourismus. Einführung in das räumlich-zeitliche System. München/Wien: Oldenbourg.

volunteerworld.com 2021: Voluntourism Opportunities. https://www.volunteerworld.com/en/volunteer-abroad/voluntourism (Zugriff 10.3.2021).

Voluntouring.org 2021: https://www.voluntouring.org/ (Zugriff 10.3.2021).

Anti-Tourism Movements

In 2017, anti-tourism activists launched campaigns against tourism or against tourists in various cities, including Barcelona, Venice, Palma de Mallorca, Amsterdam, Bhutan, and Dubrovnik (Kuščer and Mihalič 2019, p. 4). There were sometimes virulent anti-tourist protests in San Sebastián, Rome, and Dubrovnik, and in some places there were protests against cruise tourists.

After actions in Mallorca and Barcelona, there were also anti-tourist protests in the Basque Country in August 2017. Images of masked people throwing smoke bombs into a restaurant in Palma de Mallorca and holding up anti-tourism banners, or of protesters stopping a tourist bus on a street in Barcelona, spraying it and puncturing the tires, circulated on social media (Neuroth 2017). "There was neither a type of confession nor any other indication of who might be behind it. So we can not say whether it was the group 'Arran' that was responsible for similar actions," Neuroth wrote (2017).

There were also anti-tourist statements in Greece. For example, the "Greece Blog" (2017) wrote: "Even on the Greek islands—where the inhabitants have welcomed tourists of low level unthinkingly until today and thereby accepted that the product is slowly destroyed—however, voices are being heard regarding the negative consequences of mass tourism. … In Kavos on the island of Kerkyra / Corfu—because of the hordes of English drunkards one of the worst places where one can find in summer—the residents decided to set certain limits. The Kavos Culture Association announced via the widely used Daily Mail that the young English who get drunk and do nothing else are no longer welcome." One may dismiss such disgruntled remarks as isolated cases, but the fact is that in many places the positive attitude towards tourism is crumbling.

There was also resistance in smaller tourist destinations. In 2017, cultural scholar René Stettler launched a petition to limit the number of tourists and day-trippers to Rigi, a lookout point near Lucerne/Switzerland. The number of visitors should be limited to 800,000 per year, and the destination should not be "sold out" in the global cheap tour-

C. J. Jäggi, *Tourism Before, During and After Corona*,
https://doi.org/10.1007/978-3-658-39182-9_7

ism through package holidays, said the initiator. In particular, the petition demanded that the construction of a new gondola lift with 14 masts be abandoned, since the mountain is already accessible from four sides by cogwheel railways and gondola lifts. The petition "No to Rigi-Disney-World" was signed 3100 times. In 2020, a new petition was launched to regulate tourism on Rigi (Luzerner Zeitung 02.11.2019). By the deadline of the second petition on 20 July 2021, hundreds of people had already signed.

However, at least in the Corona year 2020, the tourism-critical movement ran into open doors: In this year, the frequencies of the Rigi railways fell by 43.9%, the net revenue by 34.3% and the number of employees from 238 in 2019 to 215 in 2020 (Rüegger and Z'Graggen 2021, p. 15).

Some observersbelieved that the critical view of tourism could lead to a more differentiated view and to a enrichment of the discussion around the topic of overtourism . Furthermore, "a more differentiated view of differently critical groups could enrich the discussion around the topic of overtourism and represent a further step towards a better understanding of the genesis of the (supposedly) bad reputation of touring" (Reif et al. 2019, p. 381). But is that enough? If the authorities and the tourism providers do not react more strongly to the discomfort, anti-tourist actions could increase and the rejection of tourism could grow locally. It is also to be feared that tourism will lead to a renewed outbreak of the Corona pandemic in various destinations, as was the case in Portugal and Spain in summer 2021. This could give anti-tourist movements additional influence.

References

Griechenland-Blog 2017: Noch mehr Touristen? Nein, danke! 17. August 2017. https://www. griechenland-blog.gr/2017/08/noch-mehr-touristen-nein-danke/2140466/ (Zugriff 19.7.2021).

Kuščer, Kir / Mihalič, Tanja 2019: Residents' Attitudes towards Overtourism from the Perspective of Tourism Impacts and Cooperation – The Case of Ljubljana. In: Sustainability 11/6 (2019) 1823. www.mdpi.com/journal/sustainability (Zugriff 21.2.2021).

Luzerner Zeitung 2.11.2019: Rigi: Neue Petition lanciert.

Neuroth, Oliver 2017: Tourismus in Spanien: Zerstörerischer Ansturm. In: Deutschlandfunk vom 19.7.2017. https://www.deutschlandfunk.de/tourismus-in-spanien-zerstoererischer-ansturm.1773.de.html?dram:article_id=393247 (Zugriff 19.7.2021).

Reif, Julian / Harms, Tim / Eisenstein, Bernd 2019: Tourist-Sein oder nicht Tourist-Sein? In: Zeitschrift für Tourismuswissenschaft 11/3 (2019). 381 ff. https://www.degruyter.com/document/doi/10.1515/tw-2019-0022/html (Zugriff 19.7.2021).

Rüegger, Roger / Z'Graggen, Chiara 2021: Rigi Bahnen transportieren massiv weniger Gäste. In: Luzerner Zeitung vom 14.4.2021. 15.

The Demand for Mobility

8

In 1800, people in the USA traveled an average of 50 m per day, at the beginning of the twentieth century it was 50 km per day on average (Urry 2007, pp. 3–4). However, today people do not spend significantly more time traveling—namely one hour per day. According to Urry (2007, p. 4), it also does not seem that people undertake more trips in recent times—for example, the number of interstate trips in Great Britain has remained at around 1000 per year.

In recent decades, the speed of social processes and interactions has increased massively. According to Rosa (2013, p. 20; Geissler 1999, p. 89), the speed of communication has increased by a factor of 10^7, the speed of transport by 10^2 and the speed of data processing by 10^6. This in turn means that space not only "contracts" due to higher travel speeds (Rosa 2013, p. 21), but also loses importance as an obstacle to mobility. Rosa (2013, p. 21) describes this as follows: "In this process, space loses importance in many respects for our orientation in the late modern world. Processes and processes are no longer localized, and actual places such as hotels, banks, universities and industrial plants tend to become 'non-places', that is, places without history, identity or relationship.".

Rosa has represented the technical, social and everyday acceleration in the form of a circle (Fig. 8.1).

There is no doubt that tourism, mobility and traffic are closely related. While "traffic" is more based on a traditional and technical understanding, the term mobility is broader. It means any form of physical, geographical and social movement. According to Gross (2017, p. 27), tourism and mobility or traffic are closely interwoven. There is a more pronounced dependency relationship from tourism to mobility / traffic, because tourism is hardly imaginable without local movement, but traffic and mobility do not only relate to tourists.

C. J. Jäggi, *Tourism Before, During and After Corona*, https://doi.org/10.1007/978-3-658-39182-9_8

Fig. 8.1 Mutually reinforcing
technical, social and everyday
acceleration. (Mod. after
Rosa 2013, p. 44; own
representation)

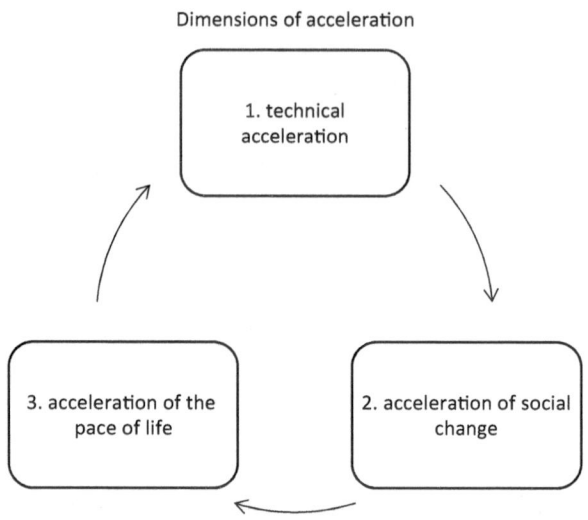

Dimensions of acceleration

Table 8.1 Different types of mobility. *Source* Gross (2017, p. 39); own representation

Mobility						
Realised mobility				Mobility as potential		
Information mobility		Social mobility		Spatial (physical, geographical) mobility		
Spiritual Mobility (intrapersonal)	Medium-bound mobility (intrapersonal)	Vertical Mobility (between the various social classes)	Horizontal mobility (within one's own social class)	(Transport) Mobility	Lifestyle mobility	Migratory mobility
Medium: immaterial (e.g. radio, internet)	Medium: material (e.g. paper)			Everyday mobility	Not everyday mobility	

Gross (2017, p. 39) has systematically presented the different forms of mobility
(Table 8.1).

8.1   Mobility, Tourism and Spaces

Mobility and tourism can also be understood as flexible or changing spatial organization.
This changes the reference frame from the "everyday space" to the "sunny beach" or the
"mountain idyll" (Wöhler 2011, p. 62). According to Wöhler (2011, p. 62), tourists move
in constructed space systems, where the constructed space is based on empirical data.
The tourist space system is zoned, that is, embedded in physical environments. Wöhler

(2011, p. 62) writes: "The tourist life in touristified spaces can also be traced back to a passage through typical space paths in concrete landscapes". For the tourist movements and flows, the individual tourist spaces have a temporary, that is, time-limited character. On the one hand, these tourist spaces are defined by non-tourist determinants, on the other hand, they are often influenced by tourism (e.g. opening times, timetables, roads, buildings, events, etc.). There is a certain contradiction between the different temporal requirements and availability, while it is expected that offers are permanently available, which is often structurally not possible. At the same time, tourism only uses these services temporarily (cf. Wöhler 2011, p. 62). So tourism goods are, so to speak, "empty goods", that is, frames or shells like physical or cultural characteristics that are present in a given contextual space. Only afterwards can it be said how these spaces are used for tourism, whereby the same space is attributed different properties or "fillings" depending on the user: for example, as a ski paradise, as a natural environment or as an event space. Tourism thus requires empty spaces and unstructured times, which are then filled with tourist offers (Wöhler 2011, p. 65). These empty spaces become tourist action spaces that are continuously reconstructed and endowed with meaning. This is the function of tourism products, which are offered as space-use rights, namely through language, through images, which help the traveler to establish a personal connection to these spaces and to occupy them for themselves (Wöhler 2011, p. 66). That is why the visualization of tourist spaces is so important.

8.2   Tourism as the Production of Resonance

Although originally an acoustic phenomenon, resonance means the phenomenon that two non-coupled bodies resonate with each other, that is, that one body's vibration stimulates the natural vibration of another body (Rosa 2016, p. 282). Therefore, sociological reciprocal adaptation movements can be understood as resonance relationships, whereby both resonate with each other. This process requires a kind of resonance space. For example, in psychoanalysis, a resonance space arises between patient and therapist, in which the therapist, through synchronous and response-resonances, takes up the situation of the patient, makes it audible and palpable, whereby the topic becomes reflexively accessible and, in the best case, processable for both (Rosa 2016, p. 286). However, resonance should not be confused with a cause-effect structure, resonance rather expresses a correlation.

Rosa (2016, pp. 341–342) speaks of horizontal resonance axes such as family, friends or (democratic) politics; of diagonal resonance axes such as the enlightening-rational relationship to objects, work or school—which he mainly sees as a resonance space (Rosa 2016, pp. 402–403)—and of vertical resonance axes such as religious reference, voice of nature, art or history.

In Rosa's opinion (2016, p. 291), resonance is only possible where people act in accordance with strong values. Because resonance is both a descriptive and a normative concept (Rosa 2016, p. 293), it practically invites tourism marketing: Tourism marketing is then most successful when it succeeds in creating a feeling of resonance with a destination or travel expectation in the person addressed.

Based on Hartwig Rosa (2016, pp. 331–332), Aschauer (2020, p. 59) has compiled resonance phenomena and -spaces in tourism (Fig. 8.2).

Aschauer (2020, pp. 59–60) distinguished the following tourist resonance spheres:

1. The body as the first resonant sphere, in particular in sports and adventure tourism;
2. the sphere of social exchange in the form of maintaining relationships with the partner, with family or relatives and friends,
3. transcendental or spiritual experiences within the framework of religious travel, relaxation offers or pilgrimage,
4. combined spheres of nature experiences, history or religion,
5. Culture as a sphere of other countries or the homeland in the form of educational travel, city trips or cultural niche offers,
6. Linking of nature and cultural experiences as a sphere of authenticity search,
7. ecological sphere in the form of ecotourism, for example by mediating a culture of slowness and idyllic nature as a contrast to the performance society.

As a particularly innovative tourism concept, Gilli and Ferrari (2016, p. 65) have referred to the Italian concept of the "extended" or "extended hotel" (*"diffuse"* or *"spread hotel"*, *"albergo diffuso"*). The individual rooms or suites—all of which offer the same

Fig. 8.2 Resonance spheres, experience spaces and suitable travel forms in tourism. (Mod. after Aschauer 2020, p. 59; own representation)

service—are located at different places, are connected by streets. The rooms or accommodation are located in different existing houses, and the occupants become a kind of local residents for a while. At the same time, tourists enjoy a more sustainable form of hospitality in the already existing rooms. The rooms are partly located in historical, traditional buildings—and the place of residence coincides with the destination—in contrast to traditional hotel tourism, where tourists stay overnight and visit certain attractions during the day (Gilli and Ferrari 2016, p. 72). One could say: The tourists go into a much more pronounced resonance with the place of residence and stay. The holiday guests are within walking distance of historical and other attractions of the city, they are less of a foreign body and the personal distance to the locals is smaller. In addition, they distribute themselves better. In 2014 there were already 82 "extended hotels" in Italy (Gilli and Ferrari 2016, p. 74).

8.3   Covid-19 and Mobility

In addition to the airlines whose planes remained largely on the ground during the first Corona wave in spring/summer 2020 and for which the entire year 2020 was a year of losses—see Jäggi (2021, pp. 88 ff.) –, high-speed trains also suffered greatly from the downturn. For example, Eurostar, the operator of the Channel Tunnel, was hit by Corona no less than four times: firstly by the decline in business travel, secondly by the uncertainty with regard to holiday travel, thirdly by travel restrictions due to national lockdowns and fourthly due to international entry hurdles (Triebe 2021, p. 24). The decline in passengers in 2020 compared to 2019 was 77%, that is, 11 million passengers less than in 2019 used the Eurotunnel, and only thanks to the still well-functioning months of January and February 2020 was the decline not even greater. While private airlines received generous state support, Eurostar only received short-time work compensation. Furthermore, the company had to turn to its owners, namely the French state railways SNCF (55%), the Belgian state railway SNCB (5%) and private financial investors (40%). The question then arose as to whether, due to the bankruptcy threat from Eurostar, the British state would soon have to step in.

After the number of flights in Europe had already declined by around 85% between February and April 2020 and had risen again to around 50% of the level at the end of January by August, they declined again between August and December 2020 and were again at around 40% of the previous year's level at the end of December/beginning of January (Enz 2021b, p. 25).

The numbers for the decline in air traffic for 2020, which Praprutitum (2021) compiled based on figures from the IATA, are impressive (Table 8.2).

According to statistics from the International Civil Aviation Organization Icao, passenger traffic on international routes decreased by 74% in the Corona year and by 50% on domestic routes (Enz 2021a, p. 26). While 4.5 billion passengers were counted in

Table 8.2 Decline in air traffic 2020. *Source* Praprutitum (2021)

Region	Demand 2020 compared to 2019	Capacities 2020 compared to 2019
Worldwide	−66.3%	−57.6%
North America	−66.0%	−51.6%
Europe	−70.0%	−62.4%
Asia-Pacific	−62.0%	−55.1%
Middle East	−73.0%	−64.5%
Latin America	−64.0%	−60.0%
Africa	−72.0%	−62.8%

2019, only 1.8 billion people boarded an airplane in 2020—a decrease of 60% (Enz 2021a, p. 26).

When deliveries of Covid-19 vaccine doses were delayed at the end of January 2021, KLM, which operates in partnership with Air France, suspended all long-haul flights on 21 January 2021, from which it had previously offered 270 per week. Only flights to South America and South Africa had been suspended earlier, after the emergence of mutated coronaviruses (Enz 2021b, p. 25). Even the cargo flights were affected by this temporary suspension of intercontinental flights. In 2020, 5000 of the 36,500 positions were cut at KLM, and another 1000 positions were to disappear in 2021. In December 2020, the Italian airline Alitalia announced that it would reduce its fleet from 104 to 52 and lay off half of its 11,000 employees—most of them on short-time work (Wysling 2020, p. 23). The airline, renamed ITA, expected a 46% decrease in global air traffic in 2021. However, Alitalia had been under state administration since 2017 and had not been profitable since 2002 (Wysling 2020, p. 23).

In mid-January 2021, budget airlines Ryanair and Easy Jet operated 96% fewer flights than a year earlier. At Lufthansa it was −86% and at Swiss −93%. Lufthansa Group CEO Carsten Spohr announced that only flights would be operated that generated a net cash flow (Enz 2021b, p. 25).

The corona crisis has also shown something else: Many airlines, especially low-cost airlines, were already insufficiently capitalized before the pandemic and had only insufficient liquidity buffers (Enz 2020, p. 19). Passengers who could not take their trip in the first wave of the corona virus sometimes had to wait months for a refund of tickets already paid in advance. Enz (2020, p. 19) said about this: "Anyone who can only stay afloat thanks to advance payments from customers does not have a stable business model". Enz—such companies should be under the close supervision of the regulatory authorities.

There had never been a time in the history of aviation in which so many states had subsidized "their" airlines independently but at about the same time. According to a study by the IATA, by summer 2020 alone, governments had spent $123 billion to keep airlines alive (Borer 2020, p. 8). Based on the volume of ticket sales, by mid-2020, the governments of the Netherlands, France, the USA and Switzerland were the deepest in their pockets. Not all subsidies were economically sensible, many of the state contributions were based on national interests (Borer 2020, p. 8). Because the subsidies were mostly spoken as loans, the debt ratio of many airlines rose to an unhealthy level.

The coronavirus crisis is also likely to have long-term consequences for rail transport. In March 2021, for the first time in Switzerland, Zurich Cantonal Bank ventured a forecast for the future development of commuter traffic on the railways after Corona. If it is assumed that home office is possible for about 41% of employees in Switzerland, the number of partial home office users will increase from 25% in 2019 to 37% in 2022, with an average of 1.5 days per week being worked at home in the future than in 2019 (Ehrbar 2021, p. 11). This would mean a total increase of 200% compared to the previous home office share. In addition, there would be a decline in face-to-face events at universities and colleges. In contrast, the presence of primary, secondary and vocational schools as well as vocational schools is expected to return to the pre-Corona level (Ehrbar 2021, p. 11). Overall, a long-term decline in commuter traffic of around 8% is expected. Rail officials also expected commuter traffic to decline in the future. The Swiss South-East Railway SOB expected a long-term decline in commuter traffic. The Postautos, which transported 25% fewer passengers in the Corona year 2020, also expected a "visibly slowed" growth (Ehrbar 2021, p. 11).

8.4 Future Developments

"Never ask a historian about the future," they say. That's basically true for all predictions—but in fact that's exactly what the prognosticators do.

Possible future developments are always difficult to predict. Predictions usually suffer from the fact that they simply continue previous trends without foreseeing sudden crashes, unanticipated changes or new causal factors. One of these crashes was the 2020 Corona pandemic.

Coghlan (2019, p. 218), based on Scott and Gössling (2015), already before Corona challenges, opportunities and risks for tourism in the next 10–20 years (Table 8.3).

Coghlan (2019, p. 73) has related the number of tourists to the acceptance of the local population (Fig. 8.3).

While many airlines kept their machines on the ground, closed borders and reduced international tourism offers, there were also voices that saw the Corona crisis positively. For example, Johannes Reck, co-founder and CEO of Getyourguide, the world's largest online booking portal for tours and attractions (Benz 2020, p. 1) explained: "We will sprint out of the crisis". But even this company was not spared from the Corona crisis—the young company had to lay off 10% of its staff and register for short-time work. But Reck said that travel would become more technological and digital after the pandemic. By the end of 2020, 80% of bookings at Getyourguide were made via the smartphone, while before the crisis it was only 40% (Benz 2020, p. 1). So, for example, the booking

Table 8.3 Opportunities and risks for tourism in the next 10–20 years. (From Coghlan 2019, p. 218, edited by the author)

Rank	Challenges	Global risks
1	Preservation of nature at tourist destinations	Fiscal crises in major economic areas
2	Balance between sustainability, environmental situation and growing markets	High structural unemployment or underemployment
3	Quick reaction to unforeseen events such as volcanic eruptions, terrorist attacks, natural disasters and crisis plans	Water disasters
4	Inclusion of green development goals in tourism	Extreme income inequality
5	Uncertainty about the impact of climate change, especially at the local level	Failure to reduce and adapt to climate change
6	Further development of sustainable tourism and corresponding management strategies in the globalized travel world	Greater likelihood of extreme weather events such as floods, storms, fires, etc.
7	Increasing investments in qualified personnel despite higher cost pressure on employers in the tourism industry	Failures in global governance
8	Introduction of tools to promote sustainable tourism	Food crises
9	Dealing with the growing cost pressure in tourism, e.g. in the transport sector or as a result of increasing global competition	Problems with financial mechanisms and financial institutions
10	Adapting tourism to the rapidly changing communication media and information platforms	Increasing political and social instability

of tickets to museums via the travel agency is outdated, and the time when tourists had to queue for hours in front of the Louvre or the Vatican is a thing of the past. Day prices would be flexible depending on the time of day and night, and the visit to overpopulated places could be indirectly controlled. In the future, all travel needs will be increasingly covered by booking platforms, including flights, car rental, tours and excursions, airport transfers, accommodation, etc.

Laura Meyer (2021, p. 3), CEO of the Swiss travel agency Hotelplan, expected that after Corona the desire to travel would remain, but the trend towards holidays closer to home and away from short trips by plane would probably lead. On the other hand, it is likely that people will enjoy longer, but less frequent, long-distance travel, and the recovery of the cruise industry could be delayed for a long time.

In contrast, in mid-2021 Matthias Benz predicted three likely developments for tourism: first, holiday tourism would return to its old tracks after Corona, with a short-term

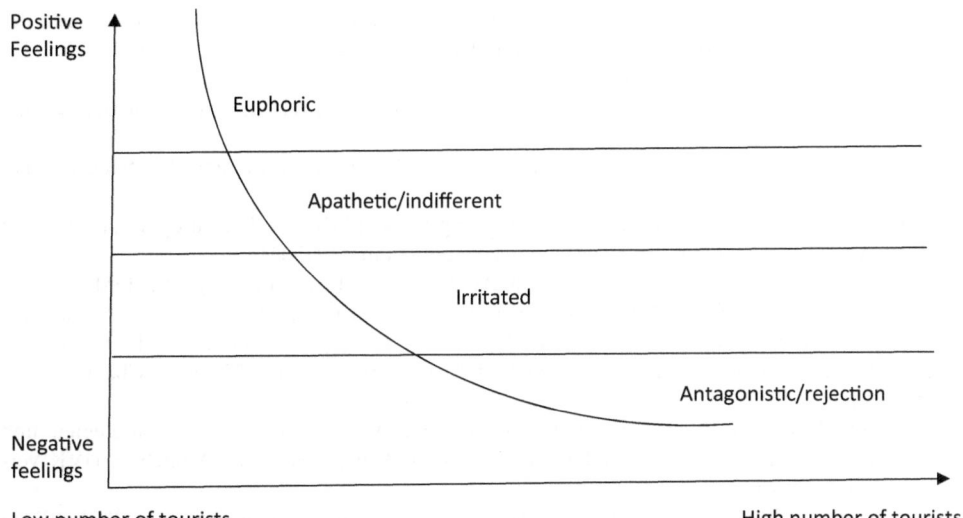

Fig. 8.3 The growing number of tourists correlates with an increasingly negative attitude of the local population. (Mod. after Coghlan 2019, p. 73; own representation)

catch-up effect. Secondly, the old tourism problems would reappear, such as climate change, overtourism and the gradual destruction of natural and cultural resources, as well as rising rents at the destination. The biggest change would be seen—thirdly—in business travel, because it had been found that virtual meetings worked just as well as personal meetings (Benz 2021, p. 17). However, I would argue against the third point that this is not certain, because many businesses still rely on personal meetings and because virtual meetings have shown during the Corona period that they cannot or only partially replace personal meetings.

References

Aschauer, Wolfgang 2020: Wenn die Welt zu uns spricht – Reisen im Zeitalter spätmoderner Entfremdung. In: Reif, Julian /Eisenstein, Bernd (Hrsg.): Tourismus und Gesellschaft. Kontakte – Konflikte – Konzepte. Schriften zu Tourismusund Freizeit. Band 24. Berlin: Erich Schmidt Verlag. 49ff.

Benz, Mattias 2020: Chancen für einen besseren Tourismus. Positive Aussichten für die Zukunft des Reisens. In: Neue Zürcher Zeitung vom 7.12.2020. 1.

Benz, Mattias 2021: Wieder frei reisen – aber anders. In: Neue Zürcher Zeitung vom 14.6.2021. 17.

Borer, Daniel 2020: Epidemie in der Luftfahrt. In: Neue Zürcher Zeitung vom 4.7.2020. 8.

Coghlan, Alexandra 2019: An Introduction to Sustainable Tourism. Oxford: Goodfellow Publishers.

Ehrbar, Stefan 2021: Fast jeder zehnte Pendler bleibt zu Hause. Nach Corona steige die Zahl der Homeofficetage um 200%, schätzt eine Bank. Die ÖV-Branche rechnet mit einer Durststrecke. In: Luzerner Zeitung vom 6.3.2021. 11.

Enz, Werner 2020: Die unsoliden Airline-Finanzen kommen ans Licht. In: Neue Zürcher Zeitung vom 12.8.2020. 19.

Enz, Werner 2021a: Die Luftfahrt steht vor einer doppelten Herausforderung. In: Neue Zürcher Zeitung vom 30.4.2021. 26.

Enz, Werner 2021b: KLM stellt kurzfristig ihre Langstreckenflüge ein. Die europäische Luftfahrt ist erneut auf dem Sinkflug. In: Neue Zürcher Zeitung vom 23.1.2021. 25.

Geissler, Karlheinz 1999: Vom Tempo der Welt. Am Ende der Uhrzeit. Freiburg/Br.: Herder.

Gilli, Monica / Ferrari, Sonia 2016: The „Diffuse Hotel": An Italian New Model of Sustainable Hospitality. In: Russo, Antonio Paulo / Richards, Greg (Hrsg.): Reinventing the Local in Tourism. Producing, Consuming and Negotiating Place. Bristol/Buffalo/Toronto: Channel View Publications. 65 ff.

Gross, Sven 2017: Handbuch Tourismus und Verkehr. Verkehrsunternehmen, Strategien und Konzepte. 2., vollständig überarbeitete und erweiterte Auflage. Konstanz/München: UVK Verlagsgesellschaft/Lucius.

Jäggi, Christian J. 2021: Die Corona-Pandemie und ihre Folgen – ökonomische, gesellschaftliche und psychologische Auswirkungen. Wiesbaden: Springer Gabler.

Meyer, Laura 2021: „Die Buchungen nehmen zu". Gespräch mit Laura Meyer, Chefin von Hotelplan. Von Benjamin Weinmann. In: Luzerner Zeitung vom 13.4.2021. 2 f.

Praprutitum, Kamolwat 2021: Turbulent Times for Travel. In: Bangkok Post vom 1.1.2021.

Rosa, Hartmut 2013: Beschleunigung und Entfremdung. Berlin: Suhrkamp.

Rosa, Hartmut 2016: Resonanz. Eine Soziologie der Weltbeziehung. Frankfurt/Main: Suhrkamp.

Scott, Daniel / Gössling, Stefan 2015: What Could the Next 40 Years Hold for Global Tourism? In: Tourism Recreation Research. Nr. 40/3 (2015). 269 ff.

Triebe, Benjamin 2021: London und Paris zanken um Hilfe für den Eurostar. In: Neue Zürcher Zeitung vom 27.1.2021. 24.

Urry, John 2007: Mobilities. Cambridge/UK / Malden/MA: Polity Press.

Wöhler, Karlheinz 2011: Touristifizierung von Räumen. Kulturwissenschaftliche und soziologische Studien zur Konstruktion von Räumen. Wiesbaden: VS Verlag für Sozialwissenschaften / Springer Fachmedien.

Wysling, Andres 2020: Alitalia wird auf die Hälfte verkleinert und erhält einen neuen Namen. In: Neue Zürcher Zeitung vom 21.12.2020. 23.

Right to Mobility?

It is argued from various sides that mobility is a basic right and should not be restricted. The Corona pandemic with its border closures, curfews and even the isolation of individual cities or regions (Jäggi 2021, pp. 112 ff.) has shown how vulnerable the freedom of movement of the individual is and how little it actually takes until the individual's mobility is restricted. Of course, this also affects tourism at its core, because tourism lives on changes of location.

9.1 Freedom and Mobility

Against the background of social developments, Aschauer (2020, p. 55) has observed in relation to egoistic identity orientation that people always invest "under the aspect of their own *success yield* in health, partnership, friendship and in their own body". In doing so, they always remain—according to Aschauer (2020, p. 55)—"hidden behind the facade of the sovereign individual" in the face of uncertainties, setbacks and personal failures. Aschauer distinguishes two types of people: on the one hand, people who have enough power for their own staging and who successfully carry their personal identity strength to the outside—so-called "surfers" on fashion trends and new developments—and on the other hand, people who are no longer able to control the uncontrolled wave movements of social developments—so-called "drifters". The "surfers" are fully integrated into the capitalist performance logic and can—at least in their own perception—largely determine their own lives. The "drifters" are increasingly forced to make far-reaching future decisions, the outcome of which is uncertain for them and their life world has become too confusing for them to structure it themselves. Transferred to tourism, this could mean that there are both tourist "surfers" and—probably the majority—tourist "drifters".

C. J. Jäggi, *Tourism Before, During and After Corona*, https://doi.org/10.1007/978-3-658-39182-9_9

Russo and Domínguez (2016, pp. 15 ff.) have shown on the basis of developments in Paris and Barcelona that in many cities a new system of local hospitality and accommodation has developed against the background of globalization, which works in addition and separately from the traditional channels of tourism. Thanks to the new media, the relationship between travel providers and tourists has changed significantly. Not only are trips bought and booked online, but tourism has become a kind of exchange and encounter lifestyle for many. Communities on the net, direct contacts with friends, like-minded people and "peers", but also direct offers from accommodation by apartment exchange, Airbnb and similar facilities not only increase mobility, but also dissolve the image of the classical tourist—which still exists, but is increasingly being challenged by other travelers. In many quarters of Barcelona, the number of "open overnight accommodation" by private individuals and in private apartments has already exceeded the number of hotel overnight stays, in Paris many quarters in terms of private accommodation, apartment exchange, etc. are on a par with the hotels or have already overtaken their offers (Russo and Domínguez 2016 , pp. 28 ff.). Even if traditional tourism still dominates overall and the phenomenon of private accommodation and peer-to-peer tourism is still concentrated on relatively few large cities—especially in view of the fact that a global middle class has emerged which has the means to travel everywhere—it is to be expected that such forms of accommodation will spread further.

9.2   Costs of Mobility

In countries like Switzerland, the majority of transport costs are borne by motorized road traffic, namely 80%. 12% of transport costs are borne by rail transport, 7% by air transport and 0.4% by shipping on rivers and lakes (Quandt and Gigon 2011, p. 42).

Of the transport costs, 59% fall on the means of transport, 14% on the infrastructure, 13% on the environment and health and 12% in accidents (Quandt and Gigon 2011, p. 43).

The cost distribution is very different for motorized road traffic and rail traffic (Figs. 9.1 and 9.2—the size of the circles expresses the different total costs).

Mobility also generates costs that are not borne by the participants in traffic, the so-called external costs. In Switzerland, these uncovered external costs of mobility amounted to CHF 13.4 billion (approx. EUR 12 billion) in 2017. They consisted of environmental, health and accident costs (cf. Table 9.1).

It is assumed that external transport costs are distributed similarly in other Central European countries.

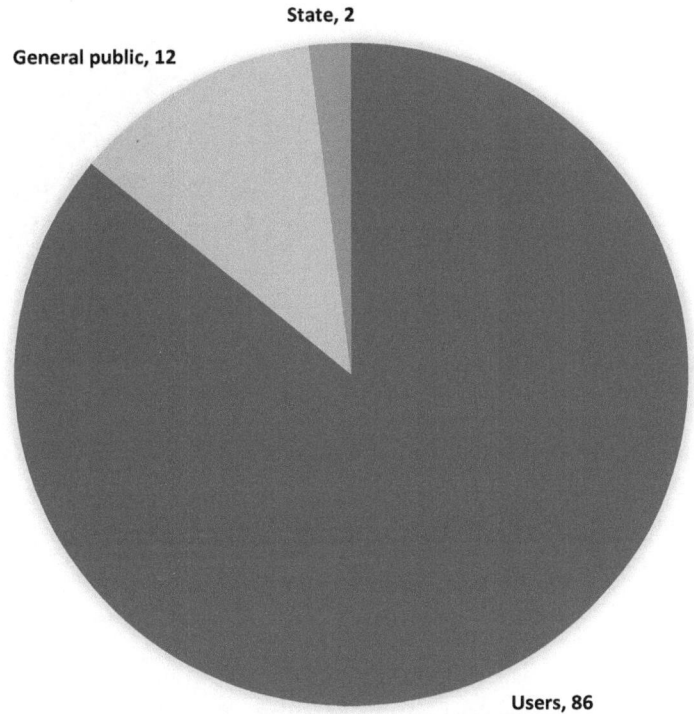

COST DRIVERS IN MOTORISED ROAD TRAFFIC IN SWITZERLAND (IN % OF TOTAL COSTS) - TOTAL 72 BN. FRANCS IN 2015

State, 2

General public, 12

Users, 86

Fig. 9.1 Cost distribution for motorized road traffic in Switzerland, a total of 72 billion Swiss francs in 2015. (Mod. after Quandt and Gigon 2019, p. 43; own representation)

9.3 Another Mobility Behavior?

Against the background of the Corona pandemic, a development has intensified that had already been discernible before: more and more people are renting a car for a short period of time or using car sharing. So a spokeswoman for a car rental company said that fewer and fewer people want to or own their own car (Ehrbar 2020, p. 9). Compared to traditional leasing, car subscriptions are cheaper, but compared to owning a car, car subscriptions are more expensive. This is shown in a comparison by the Swiss Touring Club (Table 9.2).

As can be seen from this table, the costs are approximately the same in the first six years, but when buying a car, the majority of the costs are incurred at the beginning, with the rental model spreading the costs evenly over several years, making budgeting easier. In addition, car renters can cancel their contracts with notice periods of a few

Fig. 9.2 Cost distribution for rail traffic in Switzerland, a total of 11 billion Swiss francs in 2015. (Mod. after Quandt and Gigon 2019, p. 43; own representation)

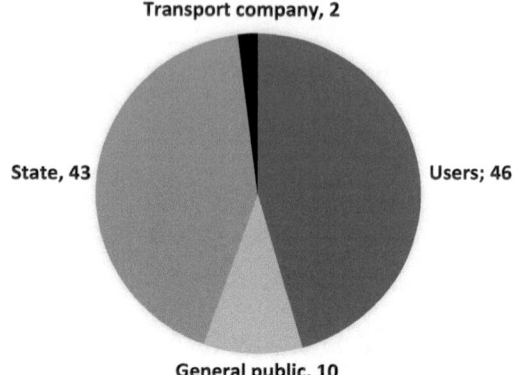

COST UNITS IN RAIL TRANSPORT IN SWITZERLAND
(IN % OF TOTAL COSTS) - TOTAL 11 BN. FRANKEN
2015

Transport company, 2

State, 43   Users; 46

General public, 10

Table 9.1 Transportation costs borne by the general public in Switzerland. (From Bundesamt für Rauentwicklung 2020)

Type of cost borne by the general public	In billion Swiss francs	Share (%)
Motorized private transport	9.5	70.9
Air traffic	1.4	10.4
Cyclists and pedestrians	1.1	8.2
Railway traffic	1.1	8.2
Public transport on the road (tram and bus)	0.27	2.0

Table 9.2 Cost comparison of car purchase and car rental. (Mod. after Ehrbar 2020, p. 9; own representation)

	Purchase		Rental	
1st year	Costs	20,000	Rent + final fee	5380 390
2nd year	Maintenance + repair	2500	Rent	5380
3. year	Maintenance + Rep	2500	Rent	5380
4. year	Maintenance + Rep	2500	Rent	5380
5. year	Maintenance + Rep	2500	Rent	5380
6. year	Maintenance + Rep	2500	Rent	5380
	Total cost	*32,500*	*Total cost*	*32,670*
7th year	Maintenance + Rep	2500	Rent	5380
	Total cost	*35,000*	*Total cost*	*38,050*

months after the minimum contract period has expired. From the seventh year on, buying is cheaper.

9.4   Changed Mobility Behavior due to Covid-19?

It is undisputed that during the Corona pandemic and in particular as a result of the tough Covid-19 measures during the various Corona waves, the mobility behaviour of people changed. For example, a tracking study by the Statistical Office of the Canton of Zurich on the whereabouts of the population aged 15 to 79 showed that mobility already decreased in October 2020 and continued to decline over the Christmas holidays (Koponen 2021, p. 11). This is interesting because there was no general lockdown in Switzerland over Christmas and New Year's Eve—the shops in Switzerland were only closed for five weeks from 18 January 2021—and that holiday shopping was allowed. In any case, the Christmas crowds stayed away in Zurich and both public transport and private cars were used less often than in previous years (Koponen 2021, p. 11). Although not comparable to the spring 2020 lockdown, the changed mobility behaviour during the second Corona wave was remarkable. The big question here was also whether this was a corona-related exceptional situation or whether a new trend was emerging.

The use of public transport has also changed through Corona. For example, in Switzerland, the Swiss Federal Railways SBB recorded a decrease in profit of almost 1 billion Swiss francs in 2019, a loss of 617 million Swiss francs was recorded for 2020. For 2021, the railway company also expected a loss of 1 billion Swiss francs. At the beginning of 2021, the number of SBB passengers was still around 50% below the pre-Corona level of 2019 (Ducrot 2021, p. 21).

Many airlines tried to counteract the decline in demand for air tickets with cheap offers. In March 2021, a flight from Switzerland to San Francisco was available for 380 francs, a flight to Bangkok for 470 francs (Weinmann and Ehrbar 2021, p. 3). Flying had rarely been so cheap. The airlines expected that the abolition of the quarantine requirement for people who have been tested or vaccinated would lead to a massive increase in bookings. Already in March 2021, when the Swiss Federal Office of Public Health removed Spain and Portugal from the quarantine list, Easyjet recorded up to 170% more ticket sales to the two countries, and Swiss announced that it would expand its flight offering over Easter 2021, as bookings to Spain and Portugal had quadrupled within a week (Weinmann and Ehrbar 2021, p. 3). However, the question remained until well into early summer whether the flights booked for summer 2021 would actually take place—especially as many countries, such as Japan or the USA, only allowed their own citizens to enter. And in many countries, entry requirements changed constantly, leading to short-term changes in flight schedules. In addition, the question arose as to how stable the point-to-point increase in demand for air travel was against the background of the situation in other world regions such as Latin America and Africa. With the possible side effects of vaccinations, continued transmission rates after vaccinations, increased risk of infection from mutations and increasing number of mutations with possibly lower vaccination protection, there remained a high degree of uncertainty and thus a high degree of unpredictability of future demand for air travel. All in all, demand and prices in air travel are likely to remain extremely volatile and therefore unstable for the foreseeable future.

References

Aschauer, Wolfgang 2020: Wenn die Welt zu uns spricht – Reisen im Zeitalter spätmoderner Entfremdung. In: Reif, Julian / Eisenstein, Bernd (Hrsg.): Tourismus und Gesellschaft. Kontakte – Konflikte – Konzepte. Schriften zu Tourismus und Freizeit. Band 24. Berlin: Erich Schmidt Verlag. 49 ff.

Bundesamt für Raumentwicklung 2020: 2017 betrugen die externen Kosten der Mobilität 13.4 Milliarden Franken. Ittigen. 18.6.2020. https://www.are.admin.ch/are/de/home/medien-und-publikationen/medienmitteilungen/medienmitteilungen-im-dienst.msg-id-79469.html (Zugriff 28.2.2021).

Ducrot, Vincent 2021: „Die Corona-Krise dürfte uns etwas mehr als zwei Milliarden Franken kosten". Gespräch mit SBB-Chef Vincent Ducrot von David Vonplon und Stefan Häberli. In: Neue Zürcher Zeitung vom 16.3.2021. 21.

Ehrbar, Stefan 2020: In der Krise mieten die Lenker ihre Autos. In: Luzerner Zeitung vom 29.7.2020. 9.

Jäggi, Christian J. 2021: Die Corona-Pandemie und ihre Folgen – ökonomische, gesellschaftliche und psychologische Auswirkungen. Wiesbaden: Springer Gabler.

Koponen, Linda 2021: So mobil war die Zürcher Bevölkerung. In: Neue Zürcher Zeitung vom 6.1.2021. 11.

Quandt, Alexandra / Gigon, Christian 2011: Die volkswirtschaftlichen Kosten des Verkehrs. In: Die Volkswirtschaft. Nr. 11 (2019). 42 ff.

Russo, Antonio Paulo /Domínguez, Alan Quaglieri 2016: The Shifting Spatial Logic of Tourism in Networked Hospitality. In: Russo, Antonio Paulo / Richards, Greg (Hrsg.): Reinventing the Local in Tourism. Producing, Consuming and Negotiating Place. Bristol/Buffalo/Toronto: Channel View Publications. 15 ff.

Weinmann, Benjamin / Ehrbar, Stefan 2021: Airlines locken mit Billigtickets – doch diese bleiben liegen. In: Luzerner Zeitung vom 18.3.2021. 3.

Solution Approaches and New Ideas

10

If it is true that tourism makes a longer-lasting contribution to the well-being of both the travelling people and the people in the host countries, it is worth considering how this effect can be amplified.

In this context, Lohmann (2019, p. 18) has listed the following effects of holidays on travellers (Fig. 10.1).

In connection with travel, Lohmann speaks of a upwards spiral of positive emotions, which eventually leads to a "memorable value" of the trip. However, one could ask whether—for example through negative travel experiences—a negative emotional spiral could not arise in the same way[1].

Lohmann (2019, p. 23) also reports, based on surveys, on the following effects of vacation travel on tourists: Tourists from all demographic groups experience happy moments on vacation, but that is by no means true for all travelers. Certain segments of tourists experience happier moments more often, such as people under 40, long-distance travelers, people on longer trips, and tourists on trips with many activities. However, such happy experiences are dependent on a minimal "readiness", that is, a "willingness to be happy", which varies greatly in practice (Lohmann 2019, p. 25).

Zenhäusern and Kadelbach (2018, p. 5) have formulated 12 theses for a tourism policy in mountain regions, which can also be applied to other tourism areas and destinations:

1. *Cooperation:* Without cooperation, the tourist destinations in the mountain region are not viable.

---

[1] I remember with horror an experience at an airport on Crete—when I had to wait for the plane for several hours in a crowded waiting room with my wife and the two small sons at that time closely packed on two square meters…

C. J. Jäggi, *Tourism Before, During and After Corona*,
https://doi.org/10.1007/978-3-658-39182-9_10

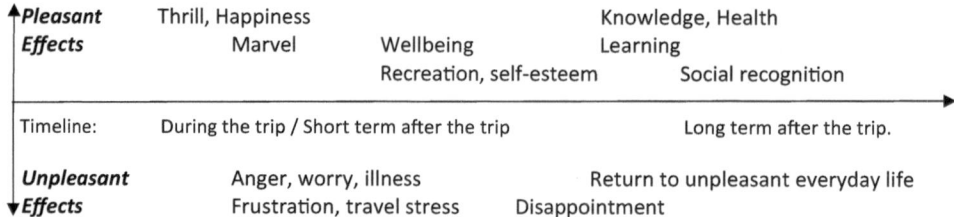

↑*Pleasant*	Thrill, Happiness		Knowledge, Health	
Effects	Marvel	Wellbeing	Learning	
		Recreation, self-esteem	Social recognition	

Timeline: During the trip / Short term after the trip   Long term after the trip.

Unpleasant   Anger, worry, illness   Return to unpleasant everyday life
▼*Effects*   Frustration, travel stress   Disappointment

Fig. 10.1 Holiday effects. (Mod. after Lohmann 2019, p. 18, own representation, slightly supplemented by the author)

2. *Price competitiveness:* Equal length skewers increase price competitiveness.
3. *Offer design:* A year-round offer secures the tourist value creation.
4. *Digitalization:* Digitalization must not remain a buzzword.
5. *Mobility:* Mountain tourism is dependent on excellent transport infrastructure.
6. *Synergies:* Cross-sectoral approaches create a win-win situation.
7. *Funding instruments:* A stronger focus of funding instruments on projects and offer design is necessary.
8. *Infrastructures:* Public investments require overarching development strategies.
9. *Tourism awareness:* The population makes a decisive contribution to tourism development.
10. *Tourist labour market:* A better anchoring of tourism in the education system and innovative employment models strengthen the tourist labour market.
11. *Sustainability:* The commitment to sustainability secures the future of mountain tourism.
12. *Regulations:* The reduction of administrative burdens lowers the cost base of tourism businesses.

However, some of these theses would have to be concretized: For example, cooperation (1) does not necessarily make sense, only if the cooperation partners "fit", digitization (4) should be customer-friendly and simplify processes, tourism awareness (9) is dependent on local conditions and very sensitive to all forms of "overtourism", sustainability (11) must be specified and operationalized—that is, made measurable and verifiable—and regulations (12) should not simply lower financial burdens, but above all also prevent the overuse of tourist resources and channel tourism into moderate channels.

10.1   Sustainable Tourism

Efforts to develop sustainable tourism have always been in the context of sustainable development concepts. Figure 10.2 shows the most important milestones in a global concept of sustainable development.

1972: 1st UN Conference on Human Development	• Explained the need to protect and improve the human environment and economic development. She reinforced the awareness that social well-being, economic development and protection of the environment were the three pillars of sustainability.
1982: World Conservation Strategy of the International Union for Conservation of Nature (IUCN)	• Strengthening the focus on biodiversity conservation.
1987: World Commission on Environment and Development (WCED)	• 'Our Common Future' report (Brundtland Report) defined sustainable development as global and intergenerational equity looking to the year 2000 and beyond.
1992: UN Conference on Environment and Development (UNCED) in Rio de Janeiro	• Main non-binding international agreements (Agenda 21, Rio Declaration, Declaration of Forest Principles) and two conventions (Framework Convention on Biological Diversity). climate change, Convention on Biological Diversity).
2002: World Summit on Sustainable Development in Johannesburg	• Declaration to achieve the Millennium Development Goals and to comply with the agreements of the International Conference on Financing for Development (Monterrey Consensus) and the Ministerial Conference of the World Trade Organisation in Doha (WTO Doha Development Agenda).
1997: Kyoto Protocol 2007: Bali Road Map 2010: Cancun Agreement (COP 16)	• Binding targets for 37 industrialised countries on greenhouse gases • Bali Action Plan with decisions for the climate future • Steps to reduce greenhouse gases
2012: UN Conference on Sustainable Development (Rio +20)	• Focus on green economy and sustainable development. Ten-year programme for sustainable consumption and production. Progress in areas such as agriculture and food security, chemicals and waste management, health and natural disasters.
2015: Climate agreement in Paris	• Agreement for the period after 2020, which obliges all countries to reduce greenhouse gas emissions. The distinction between industrialised and developing countries largely abolished.

Fig. 10.2 Milestones in a comprehensive concept of sustainable development. (Mod. after Reddy and Wilkes 2013, p. 6 and own research; own representation)

Two events were decisive for the turn to sustainable tourism. The term sustainability first appeared in the 1987 report "Our Common Future" by the World Commission on Environment and Development (Hauff 1987), also known as the Brundtland Report. The report caused international organizations and NGOs to turn to a concept of sustainable development. The Agenda 21, an action plan for the implementation of the sustainability concept, was created at the 1992 Earth Summit in Rio. Initiatives also emerged in various areas of tourism; for example, the "Mountain Agenda" studied the impacts of tourism on mountain populations and ecosystems. In 2002, the United Nations launched a "Year of the Mountains" (Keller 2017, p. 8).

McCool (2016, p. 27) has pointed out that the term "sustainable tourism" is under-stood in at least three different ways: first, as a small niche tourism offer for envi-ronmentally interested people with special needs—in particular in terms of food, accommodation and accompanying offers; second, as tourism offers particularly focused on small businesses and the local tourism industry with the aim of securing a longer-term perspective for these and third, as tourism offers focused on the entire tourism industry—also with the aim of increasing the sustainability of tourism itself and integrating tourism into the concept of sustainable development.

However, many "sustainable" tourism concepts have been more than vague. For example, the World Tourism Organization UNWTO defined sustainable tourism in 2004 as tourism in which the needs of today's tourists and host regions go hand in hand with the protection and enhancement opportunities of the future[2] (Fuchs et al. 2017, p. 14). Fuchs et al. (2017, p. 14) rightly criticized that a sustainable tourism definition must describe in detail what, for whom, with which means and to what extent it must be sus-tainable—that is exactly what the UNWTO definition of 2004 does not do.

And Mundt (2011, p. 19) already pointed out in 2011 that the term "sustainable" can be applied to almost everything as an expression of the zeitgeist. For example, Furedi (2005, p. 7) held that "sustainable" is often used as a "synonym" for "good", and any-thing that is not sustainable is considered bad. And when a politician is at a loss for words, he instinctively calls for a "sustainable something"[3].

The attitude of the tourists themselves is—to put it mildly—very ambivalent: In a study conducted in Germany in 2013, 28% of those surveyed expressed the wish that their holiday should be as ecological and socially compatible as possible, and 23% said that they would have liked to travel sustainably on their last holiday. But in practice it turned out that only 2% of the trips were based on sustainability as the central decision criterion (Hopfinger 2018, p. 22). So to put it bluntly: Sustainability yes, if it doesn't cost anything and doesn't affect the travel experience.

The tourism theorist Marco D'Eramo (2019, p. 14) even said that talking about "sustainable tourism" was about as meaningful as talking about "sustainable nuclear energy"—in other words, basically impossible. So tourism is an industry, and one that requires a solid and very heavy infrastructure.

And the situation has hardly improved. In both 2007 and 2014, travel agency employ-ees expressed the view that customers were hardly interested in the environment: The interest in environmental issues among tourists is shockingly low and no one travels to

---

[2] "Tourism that meets the needs of present tourists and host regions while protecting and enhancing opportunity for the future" (quoted from Fuchs et al. 2017, p. 14).

[3] "It appears that sustainable is good because clearly anything that is unsustainable is bad. So when politicians are lost for words their instinct is to shout 'sustainable something'" (Furedi 2005, p. 7; cf. also Mundt 2011, p. 19).

save the environment. Sustainability and environmental protection are neither a sales argument nor a decision factor (Kirstges 2020, p. 126).

In books on sustainability or sustainable tourism—such as Carnau (2011, pp. 20 ff.); Ekardt (2016, pp. 68 ff. Or Strasdas (2017, p. 14)—reference is made time and again to the sustainability triangle, according to which sustainability refers to the environment, economy and social aspects. As explained elsewhere (Jäggi 2018, pp. 31 ff.), however, the problem is that "economic sustainability", "social sustainability" and "ecological sustainability" are usually filled with completely different, even opposite, content—depending on the world view or ideology. Neoliberals talk about "sustainable development" and mean long-term profits, radical greens understand "sustainable development" as a frontal attack on capitalism, and religious circles identify sustainability with the preservation of the original creation. Ekardt (2016, p. 68) rightly criticized that, for example, profit-seeking by companies and a stable climate are formulated as equally important goals and labeled "sustainable", and hardly any political idea today does without "sustainable" and "generation-appropriate". This is especially true for the tourism sector. Ekardt (2016, p. 23) rightly points out that in the center of sustainability both the *intergenerational justice* and the *global justice* should stand.

How empty of content such platitudinous sustainability concepts can become in tourism is shown, for example, by the following example. In 2012, the "Center for Sustainable Tourism" adopted a document entitled "Pledge to Travel Green" with "Ten Ways to Care". Tourists are asked to 1.) learn about their destination, 2.) not leave good manners at home, save water and turn off the light, 3.) save gasoline by means of direct flights, only rent small cars or use the bicycle, 4.) inform themselves before making travel decisions, 5.) be a good guest and meet the local population with respect and respect their private sphere, 6.) support local craftsmen and businessmen, 7.) not throw away waste and recycle it if possible, 8.) be aware of and considerate of the environment, plants, animals and ecosystems ("mindful"), and strictly follow fire regulations, 9.) follow a zero-emission travel behavior (??) and 10.) bring sustainable behavior back home and encourage family members to travel with the same awareness (Edgell and Swanson 2013, pp. 154 f.). This all sounds very nice, but isn't it more of an etiquette guide than a plea for sustainable travel behavior?

Horrigan (2013, p. 213) has suggested that, due to the inconsistent and sometimes contradictory understanding of sustainability, this term should be redefined. Such a concept of sustainability in tourism must be based on quality and comfort on the latest technological achievements and on the benefits for the environment. This corresponds to the expectations of tourists. But the big question is whether the demanders of tourist services would accept a loss of quality and comfort, but also higher prices in favor of less negative impacts on the environment or not. If, for example, smaller and more environmentally friendly cruise ships are used, the negative impacts on site are undoubtedly lower, but also the profits of tourism providers smaller, which would most likely lead to higher prices.

Table 10.1 Core indicators for sustainable tourism management. (Mod. nach Hartmann und Stecker 2018, p. 56; leicht redigiert und ergänzt durch den Autor)

Criterion (short form)	Core indicator	Survey method	Standard
Organize	Existence of a target management organization (as a public-private partnership) pursuing sustainability goals	Secondary research, survey	Organization exists
Develop strategy	Existence of a sustainable tourism strategy for the target	Secondary research, survey	Strategy exists and is implemented
	Existence of a monitoring system for regular success control of the strategy	Secondary research, survey	System exists and is applied at least on an annual basis
Introduce quality management system	Existence of a quality management system (e.g. TQM)	Secondary research, survey	System exists, is integrated into process management and is continuously improved
Ensure safety	Existence and evaluation of a comprehensive safety and risk management system for tourists (and locals)	Secondary research, survey	System exists and is at least evaluated on an annual basis
Compliance Management	Compliance management integrated	Secondary research, survey	Compliance with laws, regulations and internal guidelines is continuously checked
Ensure accessibility	Existence of laws / regulations to support people with disabilities	Secondary research, survey	Laws / regulations exist and are implemented
	Tourist facilities and sights are barrier-free	Observation	>75% of the relevant places are barrier-free

(continued)

Table 10.1 (continued)

Criterion (short form)	Core indicator	Survey method	Standard
Carry out visitor management	Systematic monitoring of tourist numbers and seasonality	Secondary research, survey	Monitoring takes place at least on an annual basis
	Existence of a strategy to avoid/weaken seasonality	Secondary research, survey	Strategy available, operational implementation clearly recognizable, (current measures)
	Existence of guidelines to avoid overloading of tourist facilities and sights	Secondary research, survey, observation	Guidelines available, operational implementation clearly recognizable (current measures)

In addition, neither economic growth is identical to development, nor can economic growth be equated with prosperity (Mundt 2011, pp. 34 ff.). Rather, especially in the richer countries—such as in Germany—an increasing decoupling of economic growth and income can be observed. For example, real wages have been more or less stagnant for decades, despite economic growth.

Hartmann and Stecker (2018, pp. 56 ff.) have described core indicators for sustainable tourism at the *management level,* the *economic level,* the *sociocultural environment* and the *ecological sustainability level* (Tables 10.1, 10.2, 10.3 and 10.4).

Table 10.1 shows core indicators for sustainable tourism management.

Table 10.2 summarizes core indicators in the economic sector.

Table 10.3 describes core indicators for the socio-cultural environment.

And finally, Table 10.4 defines core indicators for the ecological dimension of tourism.

There is no doubt that these core indicators for the four dimensions of management, economic importance, socio-cultural interaction and environmental impact are a good and useful tool for sustainable tourism, which is not only comprehensive and applicable in practice, but also goes far beyond the usual non-binding declarations of "sustainable tourism". Even better would be if these criteria and their operationalization could be standardized so that they could be applied worldwide.

However, it should be noted that the minimum standards may differ depending on the country, region, culture and climatic and ecological environment, but it is important that measurable standards and norms are set and controlled.

Strasdas (2017, pp. 14 ff.) has proposed that, in tourism, instead of sustainability, *efficiency, sufficiency* and *consistency* be used. *Efficiency* means producing the same output with less resource use and less environmental impact, *sufficiency* means more develop-

Table 10.2 Core economic indicators. (Mod. after Hartmann and Stecker 2018, p. 57, slightly edited and supplemented by the author)

Criterion (abbreviation)	Core indicator	Survey method	Standard
Achieve local/regional economic contribution	Significant contribution to local/regional gross domestic product	Secondary research, survey	Contribution is higher than the national average (in places with high tourism intensity)
	Significant tourist spending per day	Secondary research, survey	Higher than the European average of €65 per day (in places with high tourism intensity)
Support local economic cycles	High proportion of local products in tourism-related retail and catering	Questionnaire, observation	More than half of the products come from local/regional production
Moderately increase tourist numbers	Positive development of overnight stays	Secondary research,	Slow growth rates of 1–2% per year
	Positive development of occupancy rates at accommodation establishments throughout the season	Secondary research, survey	Higher than the national average (in places with high tourism intensity) and balanced throughout the year
Generate employment effects	Share of jobs in the tourism sector	Secondary research, survey	More than 20% (based on the contribution of tourism to GDP)
	Unemployment	Secondary research	Below the national average
	Diverse training opportunities in tourism	Secondary research, survey	Specific training programs for Teaching and academic professions on site / in the region are available
Annual tourism register exists	Tourism density, tourism intensity, seasonal distribution	Secondary research	Local defined maximum values (per district/town) exist and are continuously checked

Table 10.3 Core indicators for the socio-cultural environment. (Mod. after Hartmann und Stecker 2018, p. 58; slightly edited and supplemented by the author)

Criterion (short form)	Core indicator	Survey method	Standard
Support the protection and preservation of cultural heritage sites	Existence of corresponding programs and/or concepts	Secondary research, questionnaire	Concepts/programs exist and are implemented
	Existence of a monitoring system to check for possible negative impacts of tourism	Secondary research, survey	System exists and is used at least on an annual basis
Participation: Involvement of volunteers, civic engagement (stakeholders)	Existence of a procedure for integrating local actors into decision-making processes	Survey	Procedure exists, active and regular participation is possible
Ensure the quantity and quality of employment in tourism	Existence of laws that guarantee fair wages and salaries (minority protection; minimum wage; compensation for seasonality, gender equality)	Secondary research, survey	Laws exist, all locally relevant aspects are taken into account
Ensure tourism acceptance at the destination	Degree of satisfaction of the locals with tourism	Secondary research, survey	>75% of the locals support tourism on site
Ensure visitor satisfaction	Existence of a monitoring system for guest satisfaction (e.g. perception of safety, cleanliness, noise, orientation)	Secondary research, survey	Monitoring exists and is applied at least on an annual basis
	Degree of visitor satisfaction	Secondary research, survey	>75% of tourists are fully satisfied with their stay on site
Participation of the local population	Participation opportunities for the population exist and are institutionalized	Secondary research, survey	At least annual surveys and once a year set co-decision tools
Migration review	Number of people migrating annually and seasonally in the tourism sector	Secondary research, survey	Measures to minimize fluctuation (emigration, immigration of workers, tourism flows)

Table 10.4 Core indicators for the ecological dimension of tourism. (Mod. after Hartmann und Stecker 2018, p. 59; edited and supplemented by the author)

Criterion (Short form)	Core indicator	Survey method	Standard
Climate protection: Use clean energy sources sparingly	Energy generation/-use of tourist businesses from renewable resources (water, wind, sun)	Secondary research, survey	>75% of businesses use renewable energy
	Use of energy-saving technologies and implementation of energy-saving measures in tourism businesses (accommodation, gastronomy, attractions)	Survey, observation	>75% of businesses implement measures (use of LED lights, motion sensors, key cards for rooms, etc.)
Minimizing water consumption	System for controlling water consumption	Secondary research, survey	System exists and is applied at least on an annual basis
	Daily water consumption per overnight tourist	Secondary research	<150 l per day per overnight tourist
	Use of water saving technologies and implementation of water saving measures in tourism businesses (accommodation, catering, attractions)	Survey, observation	>75% of businesses implement measures (providing information, toilet flushing, water taps)
Reduce air pollution from tourism	Existence of laws and/or guidelines to avoid greenhouse gas emissions	Secondary research	Laws/guidelines exist and are implemented
	Provision of suitable infrastructure (cycling and pedestrian paths. Bus connections, parking options)	Observation	'>75% of all tourist facilities and attractions are well accessible by bike, on foot and by bus (6–24 h); there is a dynamic parking guidance system

(continued)

Table 10.4 (continued)

Criterion (Short form)	Core indicator	Survey method	Standard
Avoid and dispose of waste properly	Existence of a modern waste system (including waste avoidance or reduction, recycling)	Secondary research, survey	System exists and is used by 75% of all tourist businesses
	Cleanliness/lack of litter in public spaces	Observation	In tourist-oriented and heavily trafficked areas, trash cans in sight
Preserve and protect biodiversity	In densely populated areas: proportion of green space	Secondary research	>10% green space in densely populated areas
Counteracting the effects of climate change	Water and wetland biotopes, trees or groups of trees providing shade in densely populated areas	Secondary research Observation	>5% water areas and >5% tree groups/ single trees
Build circular economy[4] in the tourism sector	All waste is used as raw material for other businesses	Secondary research	Approval/implementation of new businesses only as part of circular economy cluster

ment through self-restraint—for example, by foregoing luxury accommodation in favor of simple accommodation on trips—and *consistency* wants to align all offers with natural processes and cycles—for example, by using local building materials for accommodation instead of importing expensive materials. So far, only a few tourism studies have placed the lives of the local population at the center (Eilzer and Weis 2020, p. 38)—and this needs to change.

Entrepreneurial and organizational instruments for strengthening nature conservation are the definition of entry quotas, entrance and usage fees, contractual usage agreements and restrictions, and in extreme cases temporary stops for tourists.

For some time now, the development of tourist clusters—that is, coherent and interconnected tourist offers—has been discussed as a means of sustainable economic development in rural and peripheral areas. This is partly because tourism is one of the few economic sectors that specifically favors such areas (Schuhbert 2018, p. 257). National cluster systems could become diffusion media for knowledge at the global, national, local level and between centers and peripheries. Such cluster systems or networks could absorb the external and internal change pressure, increase the degree of synchronous

---

[4] For a more detailed discussion of the circular economy, see Jäggi 2021, pp. 50 ff. and Jäggi 2022, pp. 132ff.

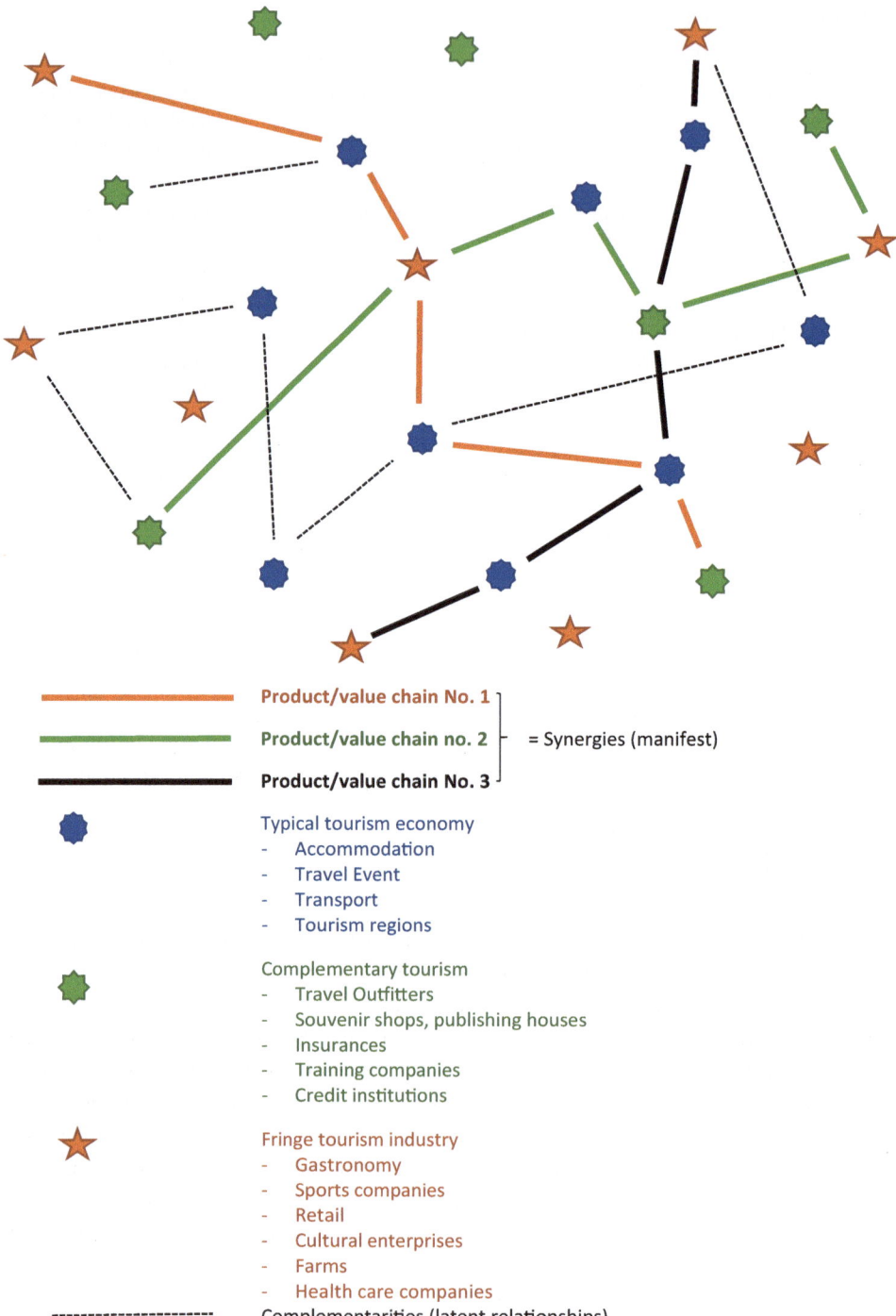

Fig. 10.3 Manifest and latent network relationships in tourist clusters. (Mod. nach Schuhbert 2018, p. 244; eigene Darstellung)

knowledge, make existing path dependencies visible and optimize them, and improve the transfer of knowledge (Schuhbert 2018, pp. 238 ff.).

Figure 10.3 shows manifest and latent network relationships in tourist clusters.

Goodwin (2016, p. 17) has pointed to the difference between sustainable and responsible tourism. Responsible travel puts the emphasis on what individuals and groups do to support sustainable travel. So it is always necessary to ask *for what* the travelers want to take responsibility, *how* they take responsibility and *what* they achieve with it. Godwin (2016, p. 147 ff.) distinguished between social responsibility, cultural responsibility, economic responsibility and environmental responsibility. However, the author remains relatively vague; So tourism should not damage the economic development of the local community, children should not be sexually or otherwise exploited, cultural heritage should be preserved and remain authentic and ecological impacts should be minimized. All this is nice and good—but how can the individual tourist influence this? For example, by inquiring about sustainability standards, discussing social, cultural and economic impacts and orienting themselves to corresponding labels. But is that enough?

In many countries, the outbreak of the Corona pandemic in tourism has mercilessly exposed how little sustainable international tourism is in many places. Dean Williams, a native New Zealander who until the Corona crisis ran one of the most prestigious gastrolocals in Siem Reap, Cambodia, described it this way: "One of our mistakes was quite clearly that we neglected the local clientele for years and instead concentrated only on travelers" (quoted after Mihai 2021, p. 10). And Williams said: "Now we understand how little sustainable this city is and how much we underestimated the danger of this massive tourism dependence in our own small bubble". The big question is, however, whether the right conclusions will be drawn from this. Dirk de Graafen, a Dutchman who ran a boutique hotel in Phnom Penh, said there were too many hotels and too many restaurants in cities like Siem Reap that all offered the same thing. There was simply too much tourism for such a small town (Mihai 2021, p. 10).

In essence, no credible concept of sustainable tourism can avoid limiting particularly environmentally damaging long-distance travel. So Christine Plüss, managing director of "Fair unterwegs", said quite clearly: "Travel means mobility, i.e. CO_2 emissions". And she recommends travelling by plane no more than every four years and staying at the destination for a longer period of time (Schaller 2019, p. 36). Gössling (2015, p. 223) named the following points for a CO_2-reducing strategy of travel:

- Reduction of the number of long-distance trips and a stronger focus on closer destinations,
- In contrast to the trend towards short stays, an extension of the duration of stay at tourist destinations,
- Change of transport means: less use of aircraft and cars, but more bus and train journeys, also avoiding cruises,
- Incentives for tourists so that they use energy-efficient transport and forms of transport, e.g. no more private planes or first-class tickets,
- Generally increased consumption of products and services that generate low CO_2 emissions or other emissions.

10.2 Unification of Environmental Labels

Similar to the area of organic or ecological food, there is also a multitude, indeed a pro-liferation of eco-labels in tourism that are hardly manageable.

For example, Amacher, Hoppler and Weber (2018, p. 66) alone for Switzerland in the year 2017 more than 20 environmental labels for hotels compiled (cf. Table 10.5).

It is clear that it is difficult or even impossible for an average tourist or even an experienced guest to keep an overview here. The travel and tourism industry should therefore do everything in its power to remedy this label chaos—unless the tourism industry is not interested in creating transparency at all …

Table 10.5 Environmental labels for hotels in Switzerland, as of June 2017. (From Amacher Hoppler and Weber 2018, p. 6)

Label	Number of hotel businesses in Switzerland with this label
QI	317
QII	151
QIII	131
Green Living	77
Sustainable Living	13
Green Key	7
Ibex Fairstay	8
ISO 14001	rd. 30
Valais Excellence	4
Green Globe	11
Bio Hotels	1
Eco Hotels Certified	4
EnAW-Energie-Modell	104
EnAW-KMU-Modell	85
Myclymate	15
Öko-Spitzenreiter (TripAdvisor)	84
Der andere Hotelführer	61
Culinarium (Ostschweiz)	8
Max Havelaar-Gastro-Partner	32
Goût Mieux	12
Travelife, Earth Check, EU-Umweltzeichen, Certified Green Hotels, Green Sign	0

10.3 Slow Tourism

The slow movement, that is, the trend towards slowing down life, has meanwhile also reached tourism. Slow tourism, like other currents of the slow philosophy, such as slow food, focuses on a change in consumption and lifestyle.

"Slow tourism" or "slow travel", that is, slow travel, means, in contrast to mass tourism, a stay or a trip to a place that, so to speak, slows down the speed of travel by, for example, immersing oneself in the local culture and nature for a week without rushing to the next place or to "sights" to be seen in the travel guide (Dickinson and Lumsdon 2013, p. 372). The aim is not to see as much as possible in the shortest possible time—in the style of "Europe in five days"—but, in contrast to the working day, to let the respective environment take effect in a "slowed down" or contemplative way.

One of the most important forms of "slow tourism" is hiking. Special forms of "slow tourism" are hikes or treks with animals, such as husky hikes, reindeer tours, donkey hikes or horse treks (Strasdas and Zeppefeld 2011, p. 63).

Figure 10.4 shows the motivation of the different age groups to go hiking.

The mobility theorist Paul Virilio had formulated the concept of dromology at the end of the 1970s, on the basis of which he set power relations and wealth of societies in correlation to speed (Döbler 2020, p. 2 as well as Virilio 2004, pp. 25 ff.). In doing so, he formulated an acceleration paradox: Thousands of vehicles with high speed potential lead to a standstill, and in the probably fastest mode of travel, flying, passengers spend a significant part of the travel time on foot in queues. Similarly, the smallest disturbance in the case of the railway leads to a standstill, because the timetables are so closely spaced and the networks are so densely used (Döbler 2020, p. 2).

For all these reasons, "Slow Travel" could be a possible answer.

Figure 10.5 summarizes the most important aspects of the tourist experience of "Slow Travel".

Thus, "Slow Tavel" or "Slow Tourism" is both a niche market and a travel model and a future market (Dickinson and Lumsdon 2013, p. 377).

The movement of Cittaslow or Slow Cities also arose from this philosophy and was founded in 1999 on the initiative of Paolo Saturnini, mayor of the Italian town of Greve in Chianti, and the mayors of Bra, Orvieto and Positano (Eilzer and Weis 2020, p. 38). The aim is to maintain and improve the quality of life in small towns, preserve local features and promote local identities. By November 2019, the Cittaslow network comprised 264 towns in 30 countries worldwide, including 21 places in Germany (Eilzer and Weis 2020, pp. 39–40). The Cittaslow initiative is still closely linked to the Slow Food movement, which it tries to extend to urban life. Criteria for membership of the Cittaslow movement include the preservation and promotion of local products and the creation of spaces for their marketing, as well as support for Slow Food activities (Eilzer and Weis 2020, p. 39).

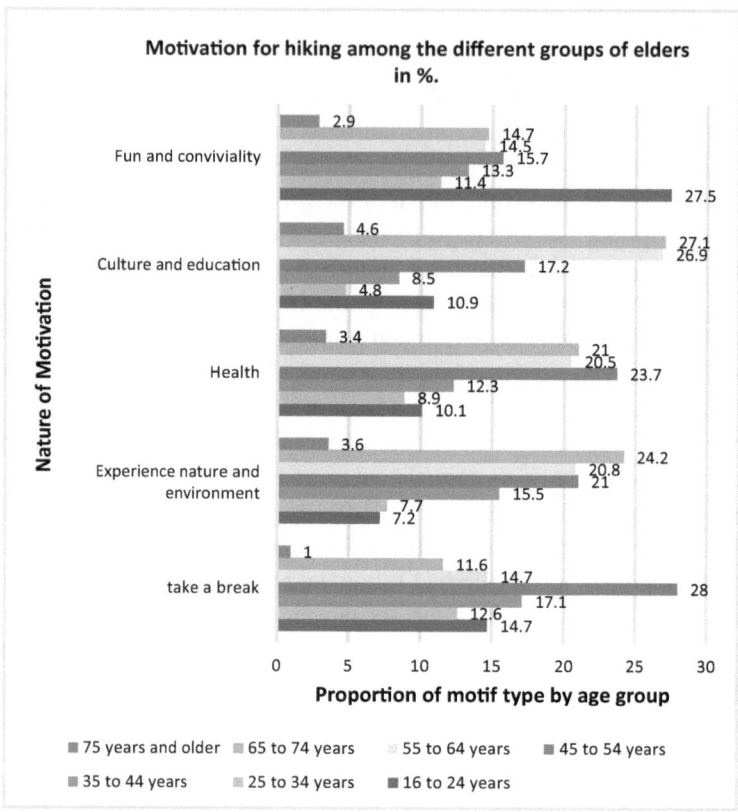

Fig. 10.4 Motivation to go hiking in the different age groups. (Mod. after Dreyer and Dürkop 2011, p. 107; own representation)

The Cittaslow movement enables towns with a population of up to 50,000 to be certified according to the following criteria: sustainable urban development, preservation of regional cityscapes and cultural landscapes, improvement of environmental quality, expansion of regional economic cycles, promotion of local features and products, strengthening of regional awareness, hospitality and international exchange, maintenance of cultural traditions and customs, quality assurance of nutrition and maintenance of traditional food culture, and strengthening of joie de vivre and quality of life (Eilzer and Weis 2020, p. 40). In summary, the Cittaslow movement can be summarized under the following three overarching goals: quality of life, deceleration and sustainability.

However, the Cittaslow movement is also subject to the well-known paradox of tourism: Positive economic effects are offset by environmental difficulties. A survey in the first Cittaslow city of Turkey, Seferihisar, showed that certification on the one hand

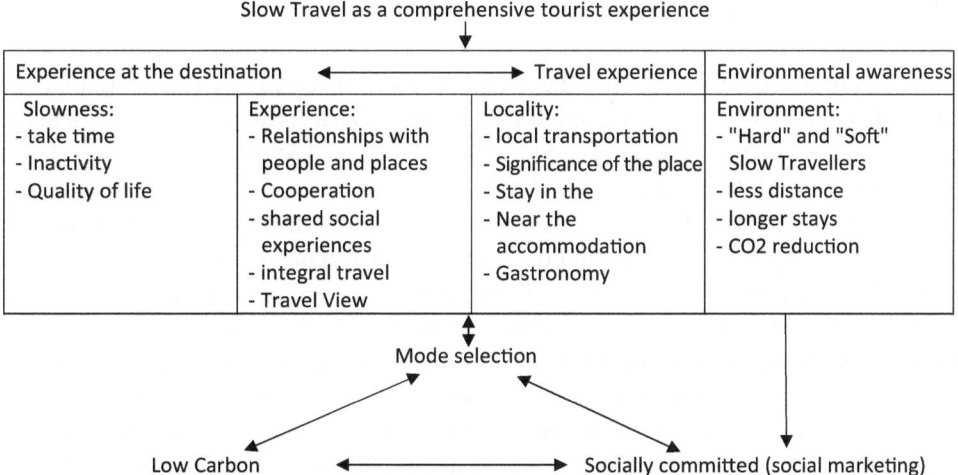

Slow Travel as a comprehensive tourist experience

Experience at the destination		Travel experience	Environmental awareness

Slowness:
- take time
- Inactivity
- Quality of life
	Experience:
	- Relationships with
	  people and places
	- Cooperation
	- shared social
	  experiences
	- integral travel
	- Travel View
		Locality:
		- local transportation
		- Significance of the place
		- Stay in the
		- Near the
		  accommodation
		- Gastronomy
			Environment:
			- "Hard" and "Soft"
			  Slow Travellers
			- less distance
			- longer stays
			- CO2 reduction

Mode selection

Low Carbon   Socially committed (social marketing)

Fig. 10.5 Slow Travel as a holistic experience of travel. (From Dickinson and Lumsdon 2013, p. 375)

improved the city's economy, cultural activities and the marketing of the city, and created additional jobs, e.g. in organic farming, but on the other hand brought more noise, traffic, numerous additional people and investors, which led to a construction boom and deterioration of the environmental situation (Eilzer and Weis 2020, p. 46). This raises the question of whether it is possible to achieve an ecological tourism through efforts such as Cittaslow.

The effects of Cittaslow were somewhat more positive in the German cities of Bad Essen, Deidesheim and Meldorf, where a study found a tendency for improved quality of life. However, the study authors themselves noted that no direct link between membership of Cittaslow and the perceived improvements could be demonstrated in the study (Eilzer and Weis 2020, p. 65). Similarly, the greater quality of life could also be a result of general social and political developments and a strengthened environmental awareness. If this is the case, Cittaslow would be little more than a new, innovative marketing label.

Slow tourism is basically a reaction of people to the increasing hectic pace of everyday life. An increasing number of people want to escape the "accelerated way of life". Although still a niche product (Cook et al. 2018, p. 391), but with growing importance, slow tourism is likely to gain further importance in the future.

10.4   Resilience in Tourism

The concept of resilience has only been associated with tourism or tourist destinations for a relatively short time (Dodds and Butler 2019, p. 2). According to these two authors, resilience in tourism can be understood as the ability of destinations to absorb shocks and disturbances (impacts) and recover[5].

Hall et al. (2018, p. 146) described resilience as a measure of how well tourism organisations, destinations, communities and individual tourists can survive, adapt and respond to increasing global and local change and unexpected disruptions[6].

Originally, the concept of resilience was also understood descriptively in tourism. In recent years, however, it has developed into a mix of normative and positive aspects that attempt to combine various system aspects of organizations and individuals as well as goals (Hall et al. 2018, p. 147). In the opinion of these authors, social systems behave differently from vulnerable environmental systems, but both are in a symbiotic relationship, which is why ecological and social resilience concepts must be integrated into tourist resilience concepts (Hall et al. 2018, p. 148). This is closely linked to change at all levels. Stability and control do not have to be the central criteria—rather, it is about adaptability in case of disruptions or crises. Resilience is a relational concept that loses its power without the concept of vulnerability and the idea of adaptability. Accordingly, resilience is significant for sustainability as well as a basis for sustainability decisions, without thereby replacing the concept of sustainability (Hall et al. 2018, p. 151). Resilience is to be seen against the background of constantly changing and developing systems as a bridge between biophysical natural sciences and social sciences. Governance issues, organizational learning and, in a wider sense, management strategies are associated with this.

10.5   Tourism Awareness

Zenhäusern and Kadelbach (2018, p. 29) have pointed to the problem that in mountain areas, but also in other tourist destinations, the awareness of tourism and the identification with the tourism industry are often not very pronounced among the local population. The success of tourism depends to a large extent on whether it is possible to increase the awareness of tourism among the population and make local people important actors in tourism. New apps and social networks could open up many new opportunities and facil-

---

[5] "the ability of destinations to absorb shocks and disturbances (impacts) and recover" (Dodds and Butler 2019, p. 2).

[6] "Resilience is significant for understanding how tourism organisations, destinations, communities and tourists as individuals can survive, adapt, respond and change in the face of increasing global and local change and disturbances" (Hall et al. 2018, p. 146).

itate the exchange between tourists and locals. For this, dialog forums are also needed, the involvement of locals and second home owners and schools (Zenhäusern and Kadelbach 2018, p. 30).

However, according to the two authors, a commitment to sustainability in tourism is also necessary (Zenhäusern and Kadelbach 2018, p. 33).

According to Réau and Guibert (2020, p. 3), tourism, if one leaves aside the commercial aspects, is about socially defined time, which has a different function depending on the social group: For lower social classes, free time is at the center, for middle and upper classes rather the leisure time loaded with cultural and educational content—among other things also to distinguish oneself from others. At least that's how it was in the early days of tourism. However, it is questionable whether—as Réau and Guibert (2020, p. 3) believe—Covid-19 is actually "a catalyst for a new leisure policy", as the two authors believe, for example in the case of China, which uses the travel behavior of several hundred million Chinese to travel abroad—including to France as the most popular European travel destination—as a geopolitical weapon. China itself would also determine what the Chinese would have to visit as tourists in their own country.

10.6   Shift to More Ecological Transport

In his interview for National Day 2020, French President Emmanuel Macron declared that he wanted to make the expansion of rail transport and freight transport by rail a priority for his last two years in office. He also linked some of his aid measures for transport companies damaged by the coronavirus with the demand for increased environmental compatibility. For example, he called on Air France to reduce domestic flights if they could be covered by rail within 2½ hours (Belz and Rasch 2020, p. 19). This is not to be underestimated, as the French railway company SNCF is 100% owned by the French state. Even if the French railways recorded a loss of 5 billion € in the first half of 2020, 4 billion of which were due to the Covid-19 epidemic and most of the rest to the strikes against the pension reform (Belz and Rasch 2020, p. 19). In addition, Macron demanded the strengthening and expansion of small, threatened railway lines in the regions, an expansion of the night train offer and an expansion of freight transport by rail (Belz and Rasch 2020, p. 19). This would mean a departure from the railway policy pursued in France for years, which primarily prioritized high-speed connections between centres at the expense of regional transport. Just two years earlier, the government had obliged the state-owned company to focus on profitable areas (Belz and Rasch 2020, p. 19).

The European Union declared 2021 the "Year of the Railway" and wanted to remind the public with campaigns that rail transport is much more environmentally friendly than other means of transport, including in terms of CO_2 emissions. According to a pan-European survey from 2017, rail transport accounted for just 0.5% of emissions, while road transport accounted for 72% and air transport for 14% . Other studies put the CO_2 emitted by the railway at 36 g per passenger kilometre, by domestic flights at 158 g and by petrol and diesel cars at 120 g per person kilometre .

The Supaéro-Décarbo group proposed making Air France a rail transport company cooperating with the French national railway SNCF, which would reduce fuel consumption. The group also called for a ban on business aircraft, a largely reduced number of domestic flights, and the dismantling of frequent flyer programs (Descamps 2020, p. 11). Other options for more climate-friendly air traffic include electrification of the ground fleet, replacement of jet aircraft with propeller aircraft on regional routes, restriction of fuel transport, and optimization of flight routes (Descamps 2020, p. 11). But all of this is unlikely to be enough to achieve the ambitious climate goals.

Perhaps a breakthrough in railway technology could be represented by the magnetic levitation technology. China and Japan are far ahead in this respect. So at the beginning of 2021, China's high-speed rail network was almost 38,000 km, which represented 26% of the total rail network. The plan was to create up to 70,000 km of high-speed connections by 2035 (Müller 2021, p. 23). Among other things, China reserved large areas for Maglev high-speed rail routes. By 2030, two sections of the maglev were to be created: one from Shanghai to Guangzhou and another from Beijing to Guangzhou, Hong Kong and Macao. While the high-speed train from Beijing to Guangzhou currently (2021) still takes around 8 h, this time is to be reduced to 3.5 h by the maglev. This development represents a danger for the airlines. By 2021, Chinese airlines had already lost many customers to high-speed trains (Müller 2021, p. 23).

10.7   Travel or Tourism Quotas?

In the future, destinations and tour operators will have no choice but to limit the number of tourists and introduce tourist quotas.

As early as 1998, the Alpine Convention CIPRA called for a restriction of ski areas in the Alpine region (Broggi and Tödter 1998), but this demand has so far been rejected by the members of the Alpine Convention, in which CIPRA has observer status (Siegrist et al. 2015, p. 49).

In the city of Zurich, the occupancy of hotels fell from an average of 80% to 10% in the second Corona wave in autumn 2020. Before Corona, around 70% of guests came from abroad, mainly from North America (Fritzsche 2020, p. 13). In a debate about increasing the city's contributions to the Zurich Tourism organization, representatives of the Greens said that instead of further promoting air traffic, rail traffic should be expanded. This also raised the question—without it being explicitly said—of a quota on air traffic and thus of tourist arrivals and departures. This is certainly not easy in cities like Zurich, after all, several tens of thousands of jobs in the hotel, restaurant and supplier industries depend on it (Fritzsche 2020, p. 13).

Because the ambitious climate goals cannot be achieved with today's state of technology in aviation, Descamps (2020, p. 11) called for a partial quota on air traffic: "One possible model is the idea of an 'individual CO_2-Allocation', which leads to a 'flight quota'. It should also allow low-income people to travel by plane within certain limits set in the fight against global warming. Beyond these limits, there would be quasi-prohibitive prices."

Long before the term "overtourism" was invented, individual places artificially restricted the number of tourists. For example, the island of Mallorca imposed a maximum number of hotel beds—435,000. Once this number was used up, no new hotels could be built. The construction of a new hotel was only possible if an old one closed. Nevertheless, the number of arrivals continued to increase (Kotowski 2019).

The Croatian city of Dubrovnik announced in 2018 that it would only allow two cruise ships to dock per day in the future and limit the number of passengers ashore to 5000 per day. Previously, up to 9000 tourists had been allowed per day (Marti and Kolly 2019, p. 22). In 2019, the Norwegian city of Bergen also had that only three ships with a maximum of 8000 people were allowed to dock daily—in 2017 four ships with a maximum of 9000 passengers had been allowed (Marti and Kolly 2019, p. 22). In Stavanger—which, like Bergen, has 200,000 inhabitants—there was also increasingly resistance to cruise ships, which polluted the air and whose all-inclusive passengers hardly contributed anything to local value creation, except perhaps by buying a few souvenirs. And especially small villages like Geiranger at the Geirangerfjord, where around 200 cruise ships arrived every year, were completely overwhelmed by this influx—not only in terms of infrastructure, but also in terms of ecological sustainability (Hermann 2018, p. 22).

10.8   Increasing the Cost of Travel?

Urs Keßler (2019, p. 34), CEO and former marketing director of the Jungfraubahn Holding, has argued that the only effective steering measure for visitor flows is the price. That is why the journey to the Jungfraujoch is more expensive in summer than in the low season. But is it really the case that visitor flows can only be steered by prices?

It was tried anyway. For example, Venice introduced a tourist tax for day-trippers, which ranged from 3 to 10 € depending on the season (Marti and Kolly 2019, p. 22).

10.9   Inclusion of Externalized Costs

A longstanding problem in tourism—as in other economic sectors—is the difficulty that the demanders do not pay for all the costs and that a significant portion of tourism costs are externalized, that is, borne by the public. These include most environmental damage—such as climate change caused by the emission of greenhouse gases, waste problems, water consumption at tourist destinations, infrastructure costs, local environmental impacts, etc. For this reason, it is also necessary for tourism that the perpetrators—that is, the tourists—take over all the externalized costs. This will make travel much more expensive, but in the long run it will be inevitable.

10.10 Worldwide Tourism Catalog as a Prerequisite for Controlling Tourism

Already, local utilization rates are being calculated at individual locations in order to define an optimum tourism density on the one hand and a sustainability limit on the other. The size of the population, the required protection of nature and the environment, and also the capacity of the local infrastructure must be taken into account (Newsome et al. 2013, p. 244).

Chen et al. (2015, p. 155) have already proposed this—as we have seen[7]—for China.

In principle, this should be done worldwide for all tourism destinations, which would have to be brought together to form a global tourism catalog and compared with the annual tourism figures. The individual regions and places could be divided into sustainability zones, by means of which the incoming tourism would already be controlled when booking, which would undoubtedly be possible without any problems via an electronic network of digital booking platforms. The individual countries could specify maximum booking numbers per region or place and time period.

10.11 New Energy Sources in Transport?

An interesting experiment recently took place in Norwegian Finnmark, an area of roughly 48,000 square meters with around 75,000 inhabitants. There are a number of scattered settlements in this area, of which only Alta in the west, Hammerfest in the north and Kirkenes in the east count more than 10,000 inhabitants. The public transport contract is carried out by small aircraft, which ensure the connection with the centers via regional airports such as Berlevag, Hasvik, Vardö and Honingsvag. It is planned that in a few years the "airbus" from Wideröe will take over the operation with smaller electric-powered propeller aircraft and even expand it. The airline wants to start operations with an eleven-seat battery-powered propeller aircraft, the P-Volt, as the first customer in 2026. 75% of the routes are to be shorter than 275 km. There are also plans at Finnair to fly to the thinly populated areas with the many regional airports in Finland and Sweden with an electric propeller aircraft. The company relies on the Swedish startup company Heart Aerospace, which is developing a 19-seat electric aircraft. According to a decla-

---

[7]Cf. Section "2.5 destinations".

ration of intent by Finnair, this airline plans to purchase up to 20 units of it (Hermann 2021, p. 24).

The Draghi government in Italy also pursued a "green" transport policy. She announced that 25 billion € would be invested in high-speed lines and freight corridors by rail, with container ships from Asia already being unloaded in Gioia Tauro in Calabria and traveling north by rail with time savings. In addition, 3.6 billion € were to be used for new, low-emission buses (Wysling 2021, p. 24).

Both Italy and Spain want to ban gasoline and diesel-powered passenger cars by 2040. And France bans domestic flights on routes that can be reached by train within two and a half hours (Brändle 2021, p. 6). This trend will undoubtedly intensify in the coming years. However, there is a danger that the increased electricity consumption of electric vehicles will lead to a "greenwashing" of nuclear energy, which is already being elevated to "green energy" by some politicians today—for example in France.

10.12   New Living Arrangements

The quarantine rules associated with Corona have also led to new living habits or strengthened new living arrangements in some places. When a newcomer from Amsterdam had to go into quarantine for a week before starting his new job on March 1, 2021, he moved into a room in the co-living "Domo Vuelo" in Kloten. A number of people of widely varying ages between about 20 and 56 years old lived in this converted hotel. Each person lived in a room of about 22 square meters. 30 min by foot from Kloten Airport and a few 100 m from the train station, people here lived together and shared a kitchen and other communal rooms. The living arrangement is less cramped than in a shared apartment, but the people have contact with each other and look out for each other. The kitchen and communal room are cleaned regularly by an external cleaning company, and each person has their own privacy in their own room with a bathroom. This form of living is marketed mainly through social media. One of the promoters of the co-living form of living said: "The biggest challenge here is to get people to participate" (quoted according to Rey 2021, p. 16). Regular community meetings are held to discuss and resolve open questions. "Conflicts are rare. Probably it helps that someone comes by once a week and cleans the kitchen" (quoted according to Rey 2021, p. 16). Expats learn German in the co-living house, natives learn English. All this seems to confirm the trend that many people who live alone in particular prefer flexible living arrangements with well-balanced privacy and moderate social contacts—perhaps a model of living for the future, especially in times of crisis or pandemics.

Different hotels have recognized the signs of (corona) times. For example, Swissôtel in Oerlikon stopped operations in autumn 2020, and the Zurich startup company Novac-Solutions GmbH set up a temporary co-living with 250 rooms there. 100 rooms were still offered as a self-check-in hotel. In X-tra, also in Zurich, rooms were converted into furnished studios on a monthly basis, and a hotel belonging to the Welcome-Hotels group set up 40 living units as "serviced appartments" (Rey and Schenkel 2021, p. 17). The Hotel Belvoir in Rüschlikon equipped part of its rooms as home office locations, which could be rented for 120 francs per day or 360 francs per week, including internet, parking and coffee. Other hotels in the vicinity of airports also set up co-living spaces (Rey and Schenkel 2021, p. 17).

10.13 Creating New, Replicated Originals as an Alternative?

The economist Bruno S. Frey (2020, p. IX) has proposed to follow a strategy of *creating new originals* instead of restricting access to known and tourist-overloaded attractions. For this purpose, he suggests using both augmented reality and virtual reality. Frey (2020, p. IX) writes about his vision: "To do this, the most important sights will be replicated identically at a new location and attractively designed using digital technology. … In this way, the history and culture of the sights will be conveyed in an exciting and instructive way. At the same time, the New Originals are easily accessible and ecologically sustainable. There will also be a suitable range of restaurants, hotels and souvenir shops". According to Frey, even new originals could offer more than existing originals, because they could be linked and networked with historical developments or other cultural events in addition.

The idea is undoubtedly future-oriented and promising. In particular, with holographic technology it will be possible to reproduce an unlimited number of three-dimensional images of locations, sights and natural impressions and to present them at any location. Long, environmentally harmful and energy-consuming journeys could thus be avoided or reduced. Virtual spaces, virtual journeys and imaginary worlds can be experienced in a realistic way. Not only nature, but also art and imagination can generate completely new experiences—even from home.

However, the question arises as to whether such a tourism-art-world (in the double sense!) will not lead to new, gigantic consumption of resources, energy and technology—for example by an enormous increase in so-called gray energy.

Because today the distinguishability of originals and copies is becoming more and more difficult, and because recent research in tourism has shown that many cultural artifacts presented as "originals" were nothing less than "authentic" and little more than copies, Frey's proposed way appears promising.

10.14 Space Flights as New Tourism?

It may seem a bit outlandish to see space travel as a replacement for today's global tourism—especially when you consider the cost and the resources required for it. But let's be honest: With almost 9 or 10 billion people on this planet, global tourism will sooner or later reach its limits, not to mention the enormous consumption of ecological resources.

On the other hand, due to digitalization, technical progress and the search for new boundaries, the broader departure to other planets is a matter of time—and as history shows, tourism always follows technical, economic or military excursions after a certain time. This should be no different with space travel.

References

Amacher Hoppler, Anna / Weber, Fabian 2018: CSR im alpinen Tourismus? Herausforderungen und Chancen. In: Mosedale, Jan / Voll, Frieder (Hrsg.): Nachhaltigkeit und Tourismus – 25 Jahre nach Rio – und jetzt? Mannheim: Verlag Metagis-Systems. 63 ff.

Belz, Nina / Rasch, Michael 2020: SNCF und Deutsche Bahn brauchen Hilfe. Die Corona-Krise lässt die Auslastung der Züge sinken – und verschärft die Probleme der Staatsbahnen in Frankreich und Deutschland. In: Neue Zürcher Zeitung vom 24.7.2020. 19.

Brändle, Stefan 2021: Frankreich sagt Inlandflügen den Kampf an. In: Luzerner Zeitung vom 4.5.2021. 6.

Broggi, M. / Tödter, U. (Hrsg). 1998: CIPRA International: Alpenreport I. Daten, Fakten, Probleme, Lösungsansätze. Bern: Haupt Verlag.

Carnau, Peter 2011: Nachhaltigkeitsethik. Normativer Gestaltungsansatz für eine global zukunftsfähige Entwicklung in Theorie und Praxis. München/Mering: Rainer Hampp Verlag.

Chen, Anze / Lu, Yunting / Ng Young C.Y. 2015: The Principles of Geotourism. Beijing/Heidelberg: Science Press/Springer.

Cook, Roy A. / Hsu, Cathy H. C. / Taylor, Lorraine L. 2018: Tourism. The Business of Hospitality and Travel. Sixth Edition. London/New York: Pearson.

D'Eramo, Marco 2019: Ist nachhaltiges Reisen überhaupt möglich? Interview mit Marco D'Eramo (Teil 3). Von Daniel Hackbarth. In: WochenZeitung vom 22.8.2019. 14.

Descamps, Philipe 2020: Luftfahrt in Turbulenzen. In: Le Monde Diplomatique (deutsche Ausgabe Schweiz). Juli 2020. 10 f.

Dickinson, Janet / Lumsdon, Les 2013: Slow Travel. In: Holden, Andrew / Fennell, David (Hrsg.): The Routledge Handbook of Tourism and the Environment. London/New York: Routledge. 371 ff.

Döbler, Katharina 2020: Die Motoren des Stillstands. In: Le Monde Diplomatique (deutschsprachige Ausgabe Schweiz). September 2020. 2.

Dodds, Rachel / Butler, Richard W. 2019: Introduction. In: Dodds, Rachel / Butler, Richard W. (Hrsg.): Overtourism. Issues, Realities and Solutions. De Gruyter Studies in Tourism. Volume 1. Berlin/Boston: De Gruyter. 1 ff.

Dreyer, Axel / Dürkop, Dorothea 2011: Slow Hiking – neue Langsamkeit im Wandertourismus? In: Antz, Christian / Eisenstein, Bernd / Eilzer, Christian (Hrsg.): Slow Tourism. Reisen zwischen Langsamkeit und Sinnlichkeit. München: Martin Meidenbauer. 105 ff.

Edgell, David L. / Swanson, Jason R. 2013: Tourism Policy and Planning. Yesterday, Today, and Tomorrow. London/New York: Routledge.

Eilzer, Christian / Weis, Rebekka 2020: Tourismus und Lebensqualität in Cittaslow-Städten – Ergebnisse einer empirischen Studie in Bad Essen, Deidesheim und Meldorf. In: Wollesen, Anja / Eilzer, Christian / Dörr, Manfred (Hrsg.): Nachhaltigkeit im Tourismus unter besonderer Berücksichtigung von kleinen Tourismusgemeinden. Herausforderungen, Implementierung, Monitoring. Ergebnisse der 3. Deidesheimer Gespräche zur Tourismuswissenschaft. Berlin: Peter Lang. 35 ff.

Ekardt, Felix 2016: Theorie der Nachhaltigkeit. Ethische, rechtliche, politische und transformative Zugänge – am Beispiel von Klimawandel, Ressourcenknappheit und Welthandel. 2., vollständig überarbeitete und aktualisierte Auflage. Baden-Baden: Nomos.

Frey, Bruno S. 2020: Venedig ist überall. Vom Übertourismus zum Neuen Original. Wiesbaden: Springer.

Fritzsche, Daniel 2020: Grüne wollen weniger Flugreisende in Zürich. In: Neue Zürcher Zeitung vom 12.11.2020. 13.

Fuchs, Matthias / Abadzhiev, Andrey / Svensson, Bo / Höpken, Wolfram 2017: A Knowledge-Based Paradigm for the Governance of Destination Sustainability. In: Pechlaner, Harald / Keller, Peter / Pichler, Sabine / Weiermair, Klaus (Hrsg.): Changing Paradigms in Sustainable Mountain Tourism Research. Problems and Perspectives. International Tourism Research and Concepts. Volume 7. Berlin: Erich Schmidt Verlag. 13 ff.

Furedi, Frank 2005: Politics and Fear. Beyond Left and Right. London/New York: Continuum.

Goodwin, Harold 2016: Responsible Tourism. Using Tourism for Sustainable Development. Second Edition. Oxford: Goodfellow Publishers.

Gössling, Stefan 2015: Carbon Management. In: Hall, C. Michael / Gössling, Stefan / Scott, Daniel (Hrsg.): The Routledge Handbook of Tourism and Sustainability. London/New York: Routledge. 221 ff.

Hall, C. Michael / Prayag, Girish / Amore, Alberto 2018: Tourism and Resilience. Individual, Organisational and Destination Perspectives. Bristol: Channel View Publications.

Hartmann, Rainer / Stecker, Bernd 2018: Nachhaltigkeitsbilanzierung im Tourismus: Entwicklung von Kernindikatoren im Städtetourismus. In: Mosedale, Jan / Voll, Frieder (Hrsg.): Nachhaltigkeit und Tourismus – 25 Jahre nach Rio – und jetzt? Mannheim: Verlag Metagis-Systems. 51 ff.

Hauff, Volker (Hrsg.) 1987: Unsere gemeinsame Zukunft. Der Brundtland-Bericht der Weltkommission für Umwelt und Entwicklung. Greven: Eggenkamp Verlag.

Hermann, Rudolf 2018: Das Kreuz mit den Kreuzfahrten In: Neue Zürcher Zeitung vom 1.11.2018. 22.

Hermann, Rudolf 2021: Emissionsfrei durch die Lüfte im hohen Norden. In: Neue Zürcher Zeitung vom 9.4.2021. 24.

Hopfinger, Hans 2018: Zwischen Scylla und Charybdis? Die Sicht der Wissenschaft auf Tourismus und Nachhaltigkeit 25 Jahre nach Rio. Ausgewählte Aspekte. In: Mosedale, Jan / Voll, Frieder (Hrsg.): Nachhaltigkeit und Tourismus – 25 Jahre nach Rio – und jetzt? Mannheim: Verlag Metagis-Systems. 15 ff.

Horrigan, David 2013: Sustaining Sustainability. In: Jenkins, In / Schröder, Roland (Hrsg.): Sustainability in Tourism. A Multidisciplinary Approach. Wiesbaden: Springer Gabler. 210 ff.

Jäggi, Christian J. 2018: Ökologische Ordnung, Nachhaltigkeit und Ethik. Problemfelder – Modelle – Lösungsansätze. Marburg: Metropolis.

Jäggi, Christian J. 2021: Säkulare und religiöse Elemente einer ökologischen und nachhaltigen Gesellschaftsordnung. EineZusammenschau. Bausteine öklogischer Ordnngen Band 5. Marburg: Metropolis.

Jäggi, Christian J. 2022: Perspektiven zum Umbau der fossilen Wirtschaft. Hürden und Chancen für nachhaltigen Konsum in Gegenwart und Zukunft. Wiesbaden: Springer Gabler.

Keller, Peter F. 2017: Changing Paradigms in Sustainable Mountain Tourism. A Critical Analysis from a Global Perspective. In: Pechlaner, Harald / Keller, Peter / Pichler, Sabine / Weiermair, Klaus (Hrsg.): Changing Paradigms in Sustainable Mountain Tourism Research. Problems and Perspectives. International Tourism Research and Concepts. Volume 7. Berlin: Erich Schmidt Verlag. 3 ff.

Kessler, Urs: „In Interlaken haben wir keinen Overtourism". Interview mit dem CEO der Jungfraubahn Holding von Nicole Tesar. In: Die Volkswirtschaft. Nr. 11 (2019). 33 ff.

Kirstges, Torsten H. 2020: Tourismus in der Kritik. Klimaschädigender Overtourism statt sauberer Industrie? München: UVK Verlag.

Kotowski, Timo 2019: Urlaubsziele vor dem Touristen-Kollaps. In: Frankfurter Allgemeine Zeitung vom 21.8.2019.

Lohmann, Martin 2019: Machen Urlaubsreisen glücklich? In: Gross, Sven / Peters, Julia Eva / Roth, Ralf / Schmude, Jürgen / Zehrer Anita (Hrsg.): Wandel im Tourismus. Internationalität, Demografie und Digitalisierung. Schriften zu Tourismus und Freizeit. Band 23. Berlin: Erich Schmidt Verlag. 15 ff.

Marti, Gian Andrea / Kolly, Marie-José 2019: Dichtestress wegen Riesenschiffen. In: Neue Zürcher Zeitung vom 5.6.2019.

McCool, Stephen F. 2016: The Changing Meanings of Sustainable Tourism. In: McCool, Stephen F. / Bosak, Keith (Hrsg.): Reframing Sustainable Tourism. Dordrecht: Springer. 13 ff.

Mihai, Silviu 2021: Menschenleere Paradiese. In: WochenZeitung vom 21.1.2021. 10.

Müller, Matthias 2021: Chinas Fluggesellschaften schwant Böses. In: Neue Zürcher Zeitung vom 1.3.2021. 23.

Mundt, Jörn W. 2011: Tourism and Sustainable Development. Reconsidering a Concept of Vague Policies. Berlin: Erich Schmidt Verlag.

Newsome, David / Moore, Susan A. / Dowling, Ross K. 2013: Natural Area Tourism. Ecology, Impacts and Management. 2nd Edition. Bristol/Buffalo/Toronto: Channel View Publications.

Réau, Betrand / Guibert, Christophe 2020: Wie geht guter Tourismus? In: Le Monde Diplomatique (deutschsprachige Ausgabe Schweiz). Juli 2020. 3.

Reddy, Vijay Maharaj / Wilkes, Keith 2013: Tourism and Sustainability: Transitions to a Green Economy. In: Vijay Reddy, Maharaj / Wilkes, Keith (Hrsg.): Tourism, Climate Change and Sustainability. London/New York: Routledge. 3 ff.

Rey, Claudia 2021: Wohnen auf 22 Quadratmetern. In: Neue Zürcher Zeitung vom 3.4.2021. 16.

Rey, Claudia / Schenkel, Lena 2021: Schliessen, umnutzen oder umdenken. Zürcher Gastbetriebe zeigen sich erfinderisch in der Pandemie. In: Neue Zürcher Zeitung vom 3.4.2021. 17.

Schaller, Zélie 2019: Nachhaltig reisen – wie geht das? In: Eine Welt. Nr. 3 (2019). 36.

Schuhbert, Arne 2018: Ländliche Regional- und Destinationsentwicklungen als Diversifikationsstrategie – am Beispiel ausgewählter Emerging Economies in Asien und Lateinamerika. In: Zeitschrift für Tourismuswissenschaft. Volume 10/Issue 2 (2018). Themenheft Internationalisierung des Tourismus – Tourismus im Wandel. Oldenbourg: De Gruyter. 233 ff.

Siegrist, Dominik / Gessner, Susanne / Ketterer Bonnelame, Lea 2015: Naturnaher Tourismus. Qualitätsstandards für sanftes Reisen in den Alpen. Bern: Haupt Verlag.

Strasdas, Wolfgang 2017: Einführung Nachhaltiger Tourismus. In: Rein, Hartmut / Strasdas, Wolfgang (Hrsg.): Nachhaltiger Tourismus. Einführung. 2., überarbeitete Auflage. Konstanz: UVK Verlagsgesellschaft. 13 ff.

Strasdas, Wolfgang / Zeppenfeld, Runa 2011: Naturtourismus und Ökotourismus. In: Antz, Christian / Eisenstein, Bernd / Eilzer, Christian (Hrsg.): Slow Tourism. Reisen zwischen Langsamkeit und Sinnlichkeit. München: Martin Meidenbauer. 55 ff.

Virilio, Paul 2004: The Paul Virilio Reader. Edited by Steve Redhead. New York: Columbia University Press.

Wysling, Andres 2021: Mauro Draghi erfindet Italien neu. In: Neue Zürcher Zeitung vom 29.4.2021. 24.

Zenhäusern, Robert / Kadelbach, Thomas 2016: 12 Thesen zur Zukunft des Tourismus in den Berggebieten. Bern: Schweizer Tourismusverband / Schweizerische Arbeitsgemeinschaft für die Berggebiete. Juli 2018.

Conclusion and Outlook

11

Given the still uncertain situation in tourism, hospitality and catering against the background of the Corona pandemic, it is very difficult to draw any conclusions from the current situation. Too many factors remain uncertain: to what extent and for how long will state aid to the tourism sector be continued, how many hotels, restaurants and transport companies will go out of business or become insolvent, how will tourist demand develop, will new Covid-19 mutations occur or will the vaccination strategy work, how will the world economy and thus the purchasing power of people develop, etc.? Furthermore, the question arises as to how tourist demand will develop in the long term and how the individual destinations will deal with mass tourism, "overtourism" and environmental problems—in particular with climate change. Will anti-tourism movements intensify? Will there be a global or national tourism destination register, will tourism quotas be introduced, will there be longer-term price increases due to environmental taxes—or will there be new restrictions due to the pandemic? How will the increasing digitalization and how will social media influence travel behavior?

A forecast of the future development of tourism has become significantly more difficult than before the Corona pandemic. But tourism will—albeit perhaps with setbacks and declines in time—continue to increase, at least as long as people have the necessary financial resources.

People's desire to travel will remain and even increase. New forms of travel and destinations—perhaps even beyond the boundaries of our planet—will be found. It will be up to the tourist actors and the politicians to develop corresponding offers, to introduce regulatory mechanisms and to steer the tourism industry in such a way that it becomes more sustainable and environmental damage and CO_2 emissions are reduced to a minimum.

© The Author(s), under exclusive license to Springer Fachmedien Wiesbaden GmbH, part of Springer Nature 2022
C. J. Jäggi, *Tourism Before, During and After Corona*,
https://doi.org/10.1007/978-3-658-39182-9_11